Sustainable Communities and Urban Housing

Since the start of the twenty-first century, urban communities have faced increasing challenges in housing affordability, with environmental issues causing additional concern. It is clear that changes to urban housing are needed to enhance the resilience of cities and improve the economic, social and physical well-being of residents. This book provides a comparative cross-national perspective on urban housing and sustainability in Europe, exploring the key barriers and drivers associated with sustainable urban development and community regeneration.

Country-specific chapters allow for easy comparison, with each first summarizing how sustainable housing operates in the country in question, before going on to discuss the key barriers and drivers at play. This book brings a sustainability perspective to the comparative housing literature which frequently fails to integrate the social, economic and environmental pillars of sustainability. The book outlines many of the changes that professionals and residents will need to make to their practices and cultures in order to enhance housing resilience. Students, researchers and professionals with an interest in sustainable community creation and regeneration will find this book an invaluable reference.

Montserrat Pareja-Eastaway is Associate Professor in the Department of Economics at the University of Barcelona (UB), Spain, where she coordinates the Research Group on Creativity, Innovation and Urban Transformation (CRIT).

Nessa Winston is Associate Professor in Social Policy at the School of Applied Social Science, University College Dublin, Ireland, where she is Director of Research.

Routledge Studies in International Real Estate

The Routledge Studies in International Real Estate series presents a forum for the presentation of academic research into international real estate issues. Books in the series are broad in their conceptual scope and reflect an inter-disciplinary approach to Real Estate as an academic discipline.

Oiling the Urban Economy
Land, labour, capital, and the state in Sekondi-Takoradi, Ghana
Franklin Obeng-Odoom

Real Estate, Construction and Economic Development in Emerging Market Economies
Edited by Raymond T. Abdulai, Franklin Obeng-Odoom, Edward Ochieng and Vida Maliene

Econometric Analyses of International Housing Markets
Rita Li and Kwong Wing Chau

Sustainable Communities and Urban Housing
A comparative European perspective
Edited by Montserrat Pareja-Eastaway and Nessa Winston

Sustainable Communities and Urban Housing

A comparative European perspective

Edited by
Montserrat Pareja-Eastaway
and Nessa Winston

Routledge
Taylor & Francis Group

LONDON AND NEW YORK

First published 2017 by Routledge

2 Park Square, Milton Park, Abingdon, Oxon, OX14 4RN

605 Third Avenue, New York, NY 10017

Routledge is an imprint of the Taylor & Francis Group, an informa business

First issued in paperback 2020

British Library Cataloguing-in-Publication Data
A catalogue record for this book is available from the British Library

Library of Congress Cataloging in Publication Data
Names: Pareja-Eastaway, Montserrat, editor. | Winston, Nessa, editor.
Title: Sustainable communities and urban housing: a comparative European perspective / edited by Montserrat Pareja Eastaway and Nessa Winston.
Description: Abingdon, Oxon; New York, NY: Routledge, 2017. |
Series: Routledge studies in international real estate | Includes bibliographical references and index.
Identifiers: LCCN 2016026060 | ISBN 9781138911482 (hardback : alk. paper) | ISBN 9781315692630 (ebook : alk. paper)
Subjects: LCSH: Housing—Europe. | Housing policy—Europe. | Sustainable urban development—Europe.
Classification: LCC HD7332.A3 S87 2017 | DDC 363.5/561094—dc23
LC record available at https://lccn.loc.gov/2016026060

ISBN: 978-1-138-91148 -2 (hbk)
ISBN: 978-0-367-73656-9 (pbk)

Typeset in Times New Roman
by Keystroke, Neville Lodge, Tettenhall, Wolverhampton

Contents

Illustrations

Figures

Photographs

Tables

Boxes

Contributors

Glen Bramley is Professor of Urban Studies at I-SPHERE (Institute for Social Policy, Housing, Environment and Real Estate), Heriot-Watt University, Edinburgh, UK, with extensive previous research experience, particularly at Bristol University's School for Advanced Urban Studies. His recent research has focused on planning for new housing, the impact of planning on the housing market, housing need and affordability, urban form and social sustainability, poverty, deprivation and the funding and outcomes of local services.

Toke Haunstrup Christensen, Ph.D., is a senior researcher at SBI, the Danish Building Research Institute, Aalborg University, Copenhagen. He studies the relationship between everyday life, energy use and sustainability with a particular focus on households and their use of technologies. His current research focuses on the integration of smart grid technologies in household practices and use of information and communication technologies in an energy perspective.

Clemens Deilmann, Prof. Dipl.-Ing., is head of Research on Resource Efficiency of Settlement Structures at Leibniz Institute of Ecological Urban and Regional Development, Dresden, Germany. In particular, his research focuses on housing demand, material flow analyses of building activities, urban structural analyses and development scenarios.

Karl-Heinz Effenberger, was Senior Researcher at Leibniz Institute of Ecological Urban and Regional Development, Dresden, Germany, until 2015. His research interests include demographic trends, housing demand and estimating housing stock over a thirty–forty-year time span.

Lars A. Engberg, Ph.D., is a senior researcher at SBI, the Danish Building Research Institute, Aalborg University, Copenhagen. His current research focuses on social housing, energy efficiency, sustainable urban development, new modes of democratic network governance, and emerging relationships between place-based policies and public service innovation. Since 2008, he has been a member of CONCITO, a Danish green think tank.

Freja Friis is a Ph.D. student at SBI, the Danish Building Research Institute, Aalborg University, Copenhagen. She holds a master's degree in Geography and Planning Studies, and her research focuses on households' everyday life practices and smart grids.

Kirsten Gram-Hanssen, Ph.D., is a professor at SBI, the Danish Building Research Institute, Aalborg University, Copenhagen. She leads the research group on Sustainable Housing and Cities and her main research interest for the past fifteen years is in sustainable housing, everyday life practices and energy consumption.

Jesper Rohr Hansen, Ph.D., is a researcher at SBI, the Danish Building Research Institute, Aalborg University. His particular research interest is in urban governance and public innovation.

Margrit Hugentobler, Ph.D., M.S.W., was recently the Director of ETH CASE (Centre for Research on Architecture, Society and the Built Environment) at ETH Zurich. Her research focuses on innovation in housing, housing for different population groups, such as women or senior citizens, multi-local living and sustainable urban development.

Jesper Ole Jensen, Ph.D., is a senior researcher at SBI, the Danish Building Research Institute, Aalborg University. His current research focuses on sustainable housing and cities.

Reinout Kleinhans is Associate Professor of Urban Regeneration and Neighbourhood Change at the Faculty of Architecture and the Built Environment, Delft University of Technology, the Netherlands. His research interests and expertise include urban regeneration, social capital, citizens' self-organization, community enterprise, collective efficacy and empowerment. He is also a board member of the Knowledge Centre Liveable Neighbourhoods in Rotterdam.

Jacob Norvig Larsen, Ph.D., is a senior researcher at SBI, the Danish Building Research Institute, Aalborg University. His research focuses on social and economic urban sustainability.

Line Valdorff Madsen is a Ph.D. student at SBI, the Danish Building Research Institute, Aalborg University. She holds a master's degree in Geography and Planning Studies and her research focuses on everyday practices related to comfort and energy consumption in housing.

Montserrat Pareja-Eastaway is Associate Professor in the Department of Economics at the University of Barcelona (UB), where she coordinates the research group on Creativity, Innovation and Urban Transformation (CRIT). Her research focuses on the analysis of urban and housing problems and policies, in particular the social and economic dimensions of these issues. Since 2000 she has been a member of the Coordination Committee of the European Network for Housing and Research (ENHR) and is currently the Vice-Chair of this network. In 2000 she established its working group on Housing and Urban Sustainability and continues to coordinate the group.

María-Teresa Sánchez-Martínez holds a Ph.D. in Applied Economics and she has been Associate Professor in the Department of Economics at the University of Granada since 1990. Her research focuses on housing finance, distributive aspects of public expenditure on housing, housing affordability, social housing, private and social rented markets and housing policies from a comparative perspective. She is a member of the research group on Creativity, Innovation and Urban Transformation at the University of Barcelona.

Eli Støa is Professor at the Department of Architectural Design and Management, Faculty of Architecture and Fine Arts, Norwegian University of Science and Technology (NTNU), Trondheim. Her main teaching and research areas are social, cultural and architectural conditions for sustainable and socially inclusive housing.

Iván Tosics is one of the principals of Metropolitan Research Institute (MRI), Budapest. He has extensive experience of urban strategic development, housing policy and EU regional policy issues. He is a Thematic Pole Manager (Programme Expert) for the URBACT programme and has been Lead Expert on two URBACT programmes. He is Vice-Chair of the European Network of Housing Research, an executive committee member of the European Urban Research Association (EURA) and Policy Editor of the journal *Urban Research and Practice*.

Catalina Turcu is Lecturer in Sustainable Development and Planning at University College London, UK. She holds a Ph.D. from the London School of Economics and previous degrees in housing, urban planning and architecture. Before entering academia she qualified and practised as an architect in the UK and Romania. Her research interests are: building/urban retrofit, housing, sustainable design and social/institutional aspects of urban sustainability.

Inga-Britt Werner is Professor Emerita at KTH, Urban and Regional Studies, Stockholm, Sweden, and a senior researcher at the department. Before her academic career, she was an architect practitioner, mainly designing housing. She completed her Ph.D. in 2000, was appointed Associate Professor in 2006 and Professor in 2011, all at KTH. Since 2004 she has been Swedish coordinator of two research education programmes: a cooperation between Swedish and East African universities. Her research focuses on perceptions of housing quality, daily life perspectives on urban form and the social development of neighbourhoods. Her ongoing research is on social change following changes in housing tenure in Stockholm neighbourhoods, commissioned by Stockholm City.

Nessa Winston is Associate Professor in Social Policy at the School of Social Policy, Social Work and Social Justice, University College Dublin. Her main research interests relate to sustainable housing, communities and regeneration. She has published extensively on these topics in both urban (sustainable urban regeneration) and rural (second and vacant homes; eco-villages) contexts. She is particularly interested in the specific needs of vulnerable social groups and communities. She is an active member of the European Network for Housing Researchers and a coordinator of its working group on Housing and Urban Sustainability.

Preface

This book would not have been possible without the input from all those who have visited the European Network for Housing Research (ENHR)'s Housing and Urban Sustainability working group . This group was established in 2000, at the request of its chair, Bengt Turner. Bengt believed that these issues should have dedicated space for discussion and exchange of ideas at the ENHR, and Montserrat Pareja-Eastaway and Eli Støa agreed to coordinate the working group. Others have shared the coordination role, most recently Nessa Winston. At the core of the working group is the view that an holistic approach to sustainable housing and urban environments is required. Over the years, the group has organized many workshops and collaborated on a range of projects on various related themes. While the composition of the group varies from year to year, some significant contributors over an extended period include Darrly Low Choy, Örjan Svane, Iván Tosics and Inga Britt Werner. We cannot mention all those who have participated and contributed at different meetings but we are most appreciative of them for contributing to our knowledge in this important area and for their inspiration. As a result, we wish to dedicate the book to the European Network for Housing Research, particularly the Housing and Urban Sustainability working group. To all of them, thank you.

Acknowledgements

Montserrat Pareja-Eastaway and Nessa Winston want to express their gratitude to the anonymous reviewers of the proposal of this book.

Montserrat Pareja-Eastaway and Nessa Winston are grateful to the University of Barcelona for enabling Dr Winston to visit that university for a period to facilitate collaboration on this book.

Nessa Winston is grateful for support in completing the book from the College of Social Sciences and Law Research Funding Scheme, University College Dublin.

Abbreviations

BER	Building Energy Rating
BREEAM	Building Research Establishment Environmental Assessment Methodology (UK)
CEC	Commission of the European Communities
CHP	Combined heat and power
CO2	Carbon dioxide
COP	Conference of Parties to UNFCCC
DGNB	Deutsche Gesellschaft für Nachhaltiges Bauen (Germany)
EC	European Commission
EDGAR	Emission Database for Global Atmospheric Research
EPBD	EU Directive on the Energy Performance of Buildings
EPC	Energy performance certification
EPI	Environmental Protection Index
ETHOS	European Typology of Homelessness and Housing Exclusion
EU	European Union
EU–SILC	EU Survey on Income and Living Conditions
GDP	Gross domestic product
GFC	Global Financial Crisis
GHG	Greenhouse gases
GINI	Measure of income (or consumption) inequality among individuals or households
HABITAT	UN Human Settlements Programme
HQE	Haute Qualité Environnementale (France)
IMF	International Monetary Fund
LA21	UN Local Agenda 21
LCA	Life-cycle cost analysis methods
LEED	Leadership in Energy and Environmental Design (US)
NGOs	Non-governmental organizations
NIMBY	Not in my back yard
NZEB	Near zero energy buildings
ODPM	Office of the Deputy Prime Minister (UK)
OECD	Organization for Economic Cooperation and Development
PPP	Purchasing power parity

SD	Sustainable development
SILC	Survey on Income and Living Conditions (EU)
UK	United Kingdom
UN	United Nations
UNFCCC	UN Framework Convention on Climate Change
WCED	World Commission on Environment and Development
WHO	World Health Organization
ZEB	Zero emissions buildings

1   Introduction

Nessa Winston and
Montserrat Pareja-Eastaway

Introduction

Urban communities are facing a range of social, economic and environmental pressures many of which fall under the broad rubric of 'sustainability'. Cities and other large urban areas are increasingly recognised as being crucial in addressing climate change and low carbon transitions (Betsill and Bulkeley, 2007; Bulkeley, 2010: Bulkeley *et al.*, 2013). While climate change effects and those relating to other environmental problems vary depending on the location, they may involve increased risks for some communities, from flooding and drought, heatwaves, energy poverty, and transport poverty. Urban housing has become, or has the potential to become, a cornerstone of strategies to achieve greater sustainability and for climate change adaptation and mitigation. From a housing perspective, there are also very particular socio-demographic challenges which need to be addressed, in particular those related to population growth and aging, and a range of approaches are required to address these issues (see, for example, Cisneros *et al.*, 2012). Transnational migration also demands implementing appropriate and adequate responses to meeting housing needs in some areas and to facilitate social integration and social cohesion. The economic and housing booms at the beginning of the twenty-first century had a range of impacts on housing systems and on many urban households. These include: housing affordability and household indebtedness; poor housing construction/regeneration; urban sprawl in some locations, densification in others; and long-distance commuting. In some cases, these were issues of concern prior to the boom but continued or accelerated during that period. While the Global Financial Crisis (GFC) had different effects in different countries, in certain locations there were increases in social exclusion, the financial vulnerability of households and rising homelessness. Other problems can be identified in specific regions, such as high levels of vacant dwellings, and the collapse of the construction industry.

This book provides a comparative cross-national perspective on urban housing and sustainability in eleven European countries (Denmark, Germany, Hungary, Ireland, Norway, Romania, Spain, Sweden, Switzerland, the Netherlands and the UK). There is a particular focus on the effectiveness of current policies and practices in promoting urban sustainability, which may be related to the broader political, economic and environmental contexts. However, Bulkeley *et al.* (2013: 5) argue that in analysing low carbon transitions it is important to pay attention to the ways in which these contexts are 'mediated by everyday life and the myriad power relations that sustain and constrain such actions'. This book explores the range of barriers to and drivers of more sustainable urban development in each of the countries under examination. In many locations, considerable regeneration of the existing built environment and neighbourhoods will be required if more sustainable communities are to be developed. It will entail the regeneration of the physical aspect (housing, public space, etc.) using environmental criteria, incorporating blue-green infrastructure (trees, pocket parks, green roofs and walls, rain ponds, etc.) and improving the natural environment. However, it also means improving the economic, social and physical well-being of residents. Frequently, it will require changing the practices and cultures of many of the relevant professionals (e.g. urban planners, the construction industry and architects, housing providers, such as local authorities and housing associations, financial institutions and utility companies) and of the residents of urban communities.

It has been argued that much can be learned from European towns and cities about how to create sustainable urban neighbourhoods (Rudlin and Falk, 2009). This book brings a cross-national, sustainable development perspective to the comparative housing literature. It draws on a range of high-quality quantitative and, in some cases, qualitative data. As a result, each chapter is both methodologically rigorous and contextually sensitive. Each of the country chapters attempts to integrate the social, economic and environmental dimensions of sustainability into its discussion of urban housing. In sum, the book places housing issues at the centre of discussions of sustainable urban development in each national context.

Conceptualising sustainable communities and sustainable housing

The terms 'sustainable communities' and 'sustainable housing' are frequently linked to the literature on sustainable development (hereafter SD). The World Commission on Environment and Development (Brundtland Report) defined 'sustainable development' as 'development that meets the needs of the present without compromising the ability of future generations to meet their needs' (WCED, 1987: 43). This is one of the most frequently cited definitions of SD and has been adopted in many UN, EU, national and local policy documents. It emphasises the entitlement of both present and future generations to a fully functioning 'common good'. It views SD as requiring three mutually reinforcing pillars: economic; social; and environmental. Others add a governance/institutional pillar (Pareja-Eastaway and Støa, 2004). Indeed, some argue that new forms of governance may be crucial for low carbon transitions (Späth and Rohracher, 2013), that different actors may lead these transitions, including, but not always, politicians, entrepreneurs and knowledge makers, and that the scope for urban innovation to have a wide impact depends on the multilevel governance context. At the core of the SD model are principles of justice, equity and limits to growth. However, critics of the model argue that it does not sufficiently emphasise the limits to economic growth in environmental and social terms and/or that 'de-growth', reducing production and consumption, is required (Schneider *et al.*, 2010; Kallis, 2011).

While there are numerous definitions of 'sustainable communities', Turcu (2009: 41) points out that 'they have been defined as an aggregate of characteristics including among others economic security and growth, environmental quality and integrity, social cohesion and quality of life, empowerment and governance'. One of the most frequently cited definitions is that contained in the Bristol Accord, the conclusions of a Ministerial Informal on Sustainable Communities in Europe during the UK's presidency of the EU in 2005 (ODPM, 2005). Drawing on the Brundtland vision of SD, the Bristol Accord defined 'sustainable communities' as:

> places where people want to live and work, now and in the future. They meet the diverse needs of existing and future residents, are sensitive to their environment, and contribute to a high quality of life. They are safe and inclusive, well planned, built and run, and offer equality of opportunity and good services for all.
>
> (ODPM, 2005: 6–7)

Social and economic dimensions of sustainable communities include the affordability of the accommodation, and residents' perceptions of its quality and of the neighbourhood in which it is located. Space for social interaction and community development may be significant. For example, it is argued that sustainable housing should provide access within walking distance to facilities such as community centres and leisure facilities (Worpole, 2003; Worpole and Knox, 2007). Finally, it is important to note that social, economic and environmental risks and vulnerabilities may be more prominent in certain communities and in certain households living within them. Beck (1992) highlights a concern for socially disadvantaged groups, noting that those most likely to experience risk are least likely to have the resources to deal with it.

Research on sustainable housing often focuses on: its location; the construction and/or design; and dwelling use (Winston, 2007, 2010). A key concern is with the ecological limits to some of the inputs to housing (e.g. land and non-renewable construction materials) and with the emissions associated with housing and the residential sector (see, for example, Huby, 1998, 2002; European Environment Agency, 2011). In terms of location, sustainable land-use planning is required which entails a shift towards more housing being constructed within mixed use developments, resisting scattered settlements and a preference for brown-field rather than green-field sites (Wheeler, 2004); proximity to good-quality public transport, which is linked to centres of employment, services and facilities (Lock, 2000; Stead, 2000; Wheeler, 2004), high-quality dwellings and neighbourhoods (Edwards and Turrent, 2000: 9–10); and a shift away from the low residential densities and built forms that are associated with standard suburban housing

(Norman *et al.*, 2006; Stead, 2000). It requires ensuring building and design practices that give rise to high-level energy efficiency in dwellings, reduce the use of non-renewable materials, utilise local sources of renewable materials, and facilitate the recycling of valuable resources such as water, energy and waste.

Indicators of sustainable communities and sustainable housing

How sustainable communities and housing are conceptualised affects the indicators which are developed to evaluate and monitor them. Work on the development of such indicators is still rather limited, a situation highlighted by Winston and Pareja-Eastaway (2008). The literature reviewed in the previous section highlights a range of relevant issues and themes from which indicators might be developed, namely: mixed used developments; residential density and built form; sustainable construction and design methods; housing affordability; housing and neighbourhood quality; energy; poverty and social inclusion; food security; sustainable transportation; recycling; and place attachment. Using data on some of these topics available in the European Quality of Life Survey, 2007, plus cross-national data on greenhouse gas emissions and renewable energy use, Winston (2014) ranked European countries in terms of the sustainability of their urban communities, revealing signification variations by location. Some of the authors in this volume draw on that work in their chapters.

Eurostat, the statistical agency of the EU, now provides data which can be used as indicators for many of the elements of sustainable housing listed in the previous section, most of which are drawn from the EU Survey on Income and Living Conditions (SILC). Table 1.1 presents some of these data for the countries examined in this volume. While recognising their limitations (e.g. they are an incomplete set and some individual indicators might be improved), they provide some relevant statistical background on the issues of interest in this volume and are used in some of the chapters. Their main advantages are that they attempt to capture the status of a country relative to others and its progress along some important dimensions; and that they attempt to tap into each of the social, economic and environmental elements of urban housing and communities. They are: the housing cost overburden rate; rent/mortgage arrears; built form (detached, semi-detached, apartment); housing quality; overcrowding; public transportation; environmental protection; poverty risk; and unemployment. The table also includes each country's score on the Environmental Performance Index (EPI), which is produced via a collaborative project between Yale and Columbia universities. There are substantial differences between jurisdictions in the number of environmental instruments they have adopted, the timing of their adoption, the stringency of regulations and their development over time (Knill *et al.*, 2012).

Sustainable communities: supranational governance

Baker (2015) has examined the importance of international diplomacy and governance in promoting sustainable development. At the international level, the

Table 1.1 Indicators of sustainable housing

Country	Housing cost overburden rate (%)					Rent / mortgage arrears (%) 2013	Dwelling type (%)			Housing quality index	Overcrowding rate	Modal split of passenger transport (train, bus) (%)	Environmental Protection Index score 2014	At risk of poverty or social exclusion %	Unemployment % 2014
	Total	Own, no mortgage or housing loan	Own, with mortgage or loan	Tenant, rent at market price	Tenant, reduced rent or free		Detached	Semi Detached	Flat						
Denmark	18.2	8.5	9.6	33.9	50.4	3.4	57.1	12.5	29.9	78.2	7.4	19.8	76.92	19	6.5
Germany	16.6	10.2	11.9	23.7	19.4	2.1	28.6	16.7	53.2	83.4	6.6	14.7	80.47	19.6	5
Hungary	13.5	7.4	28.1	38.9	19.3	2.3	63.9	5.4	30.1	71.8	47.2	32.3	70.28	32.4	7.3
Ireland	6.6	2.8	3.9	21.7	6.3	12.0	36.2	59	4.7	80.2	3.2	17.2	74.67	30	11.1
Netherlands	14.4	3.8	13	19.7	0	3.5	16.2	60	18.6	81	2.5	11.8	77.75	15	6.5
Norway	9.7	3.9	8.5	29.8	13.9	4.3	60.7	20.2	13.3	88.3	5.6	10.3	78.04	13.7	3.7
Romania	16.5	15.7	42.1	76.3	21.4	0.8	60.5	1.7	37.8	56.5	51.6	17.8	50.52	41.7	6.7
Spain	14.3	4.3	15	50.4	13	6.4	13.6	21.2	65	85.1	5.6	19.3	79.79	28.2	24.1
Sweden	7.6	8.1	3.1	16.7	20.3	2.3	50.6	8.9	40.2	87	10.8	15.8	78.09	15.6	7.7
Switzerland	12.1	8.8	6.7	16.6	10.1	2.4	23.7	13.1	59.7	81.5	5.9	22.3	87.67	17.5	3.2
UK	7.3	1.7	4.7	23.8	7.6	4.2	23.9	60.9	14.5	78.6	7	14	77.35	24.1	5.9
EU28	11.2	6.7	8.3	26.2	11.6	4.2	34.1	24	41.3	79.5	17	16.6	72.36	24.8	10

Sources: Eurostat (SILC, 2012, unless otherwise indicated); Environmental Performance Index (EPI).

Notes: Housing cost overburden rate = percentage of population living in a household where total housing costs (net of housing allowances) > 40 per cent of total disposable household income (net of housing allowances). Housing quality index = percentage of population with none of following problems: leaking roof, damp walls/floors/foundation, or rot in window frames or floor; lack of bath or shower in the dwelling; lack of indoor flushing toilet for sole use of the household; too dark/not enough light. Overcrowding rate = percentage of population living in an overcrowded household. A person is considered as living in an overcrowded household if the household does not have at its disposal a minimum number of rooms equal to: one room for the household; one room per couple in the household; one room for each single person aged 18 or more; one room per pair of single people of the same gender between 12 and 17 years of age; one room for each single person between 12 and 17 years of age and not included in the previous category; one room per pair of children under 12 years of age. EPI is an aggregate of 20 indicators of national-level environmental data. It consists of two overarching objectives: environmental health and ecosystem vitality. The former measures the protection of human health from environmental harm while the latter measures ecosystem protection and resource management. These two objectives are subdivided into nine categories covering high-priority environmental policy issues, including air quality, forests, fisheries, and climate and energy. Higher scores indicate better performance and vice versa. The EPI is a collaborative project of the Yale Center for Environmental Law and Policy (YCELP) and the Center for International Earth Science Information Network at Columbia University, in conjunction with the World Economic Forum. Harmonised unemployment rates (percentage), monthly (M09), seasonally adjusted.

UN has played a role via the World Commission on Environment and Development report, and its vision has impacted on policy in many countries. Similarly, the UN Conference on Environment and Development in Rio, 1992, led to the production of national sustainable development strategies and Local Agenda/Action 21 in many countries and local authorities. In most cases such non-binding policies are relatively limited in their capacity to effect significant change due to, for example, limited resources and lack of monitoring (OECD, 2009; Swanson *et al.*, 2004; Meadowcroft, 2007). Legally binding international agreements/directives, such as the UN Framework Convention on Climate Change's Kyoto Protocol or EU directives, are more promising, although not without implementation problems (Baker, 2015; Jordan and Tosun, 2013).

Access to housing and housing assistance are recognised in the EU's Charter of Fundamental Rights and in its European Social Charter. There is no formal EU housing policy, but its influence on EU member states has increased in a number of ways. First, there is a wide range of EU regulations on the environment which can affect housing development, such as those for water quality and flood risk (Jordan and Adelle, 2013) but the most relevant relate to the energy efficiency of dwellings. For instance, the Energy Performance of Buildings Directive required member states to: enhance their building regulations; introduce energy certification schemes for buildings; implement inspections of boilers and air-conditioners; and move towards new and retrofitted 'near zero-energy buildings' (NZEB) by 2020 (2018 for public buildings). This directive was adopted to contribute to the EU's Kyoto commitment. In addition, the Mortgage Credit Directive (2014/17) has created an EU-wide mortgage credit market with a number of provisions. These include consumer information requirements, rules and standards for the performance of services, a consumer creditworthiness assessment obligation, provisions on foreign currency loans, provisions on tying practices, financial education, arrears and foreclosures. There are, however, considerable national variations in the implementation of various EU regulations (CEC, 2015; Jordan and Tosun, 2013).

Second, a number of EU funding programmes provide finance for certain aspects of housing in member states (e.g. the EU Fund for Strategic Investments, the European Regional Development Fund and the European Social Fund). These include energy efficiency measures (e.g. social housing energy retrofitting) and housing related urban regeneration social infrastructure projects (Housing Europe, 2015). The dominance of energy in these measures and directives reflects the EU's energy and climate targets for 2030 (CEC, 2013).

Third, the intensified EU economic governance process, introduced following the GFC, involves monitoring economic trends and policies, and the European Commission has made a number of recommendations on housing markets and policies in some countries (Housing Europe, 2015).

Finally, regulations governing competition, the internal market and state aid have affected housing policy (social housing in particular) in several countries. However, the response to advice from the European Commission varies from one country to another, and Gruis and Elsinga (2014) argue that these EU effects should not be exaggerated.

Overview of the volume

This book examines the sustainability of urban housing and communities in eleven European countries. The countries were selected to reflect diversity on a range of issues but, in particular, on housing systems and welfare regimes as well as environmental performance. To enable a more holistic approach to the topic, the authors are from different disciplines: architecture; economics; political science; social policy; sociology; and urban planning. Table 1.2 provides some contextual information on each of the countries being examined, including key social, demographic and economic indicators. The countries vary significantly with regard to per capita GDP. Norway is the wealthiest on this indicator while Romania is the poorest by some considerable margin. The countries also differ in terms of the severity of impacts of the GFC. Table 1.2 reveals that unemployment in the countries under examination ranges from a high of 24.1 per cent in Spain to a low of 3.2 per cent in Switzerland. The risk of poverty and social exclusion differs significantly in these countries. Romania had the highest risk of poverty (41.7 per cent) among the countries examined while Switzerland had the lowest risk (17.5 per cent). Households may be more or less protected from poverty and some of the shocks associated with the GFC by the presence and levels of generosity of a welfare state. The book includes countries which are typically classified in a range of welfare regimes, including social democratic (Denmark, Sweden and Norway), liberal (Ireland, UK and Switzerland), Southern European (Spain), transition (Hungary and Romania) and some hybrid models (Germany – conservative/corporatist; Netherlands – social democratic/corporatist). It was essential to include a variety of housing systems, given their effect on some of the key issues of relevance in this book, namely: housing tenure, provision, and the availability of social and affordable housing. Each of the main housing systems is covered in the book, including dual (UK, Ireland) and unitary/integrated rental models (Sweden, Denmark) (see Kemeny, 1995; and Kemeny *et al.*, 2005), but also Southern European housing systems (Spain), and two from the Eastern European 'super' home ownership model (Hungary and Romania). Finally, environmental performance varies significantly across the countries, as measured by the Yale–Columbia Environmental Performance Index, which is described under Table 1.1 above. Switzerland scores highest for environmental performance (87.7) while Romania has the lowest score (50.5).

Each of the country chapters (Chapters 2–12) follows a similar format. They commence with an introduction providing some relevant background on the key economic, social and environmental issues facing the country, including the impacts of the GFC. Next, they provide some important information on the housing system and housing provision. They proceed with an analysis of sustainable housing/community policies, definitions and indicators used in the country in question. This is followed by a discussion of the main barriers to and drivers of sustainable community development and regeneration. In the final section of each country chapter, the authors present an analysis of one issue which is of particular significance or interest in their own national context.

Table 1.2 Selected characteristics of countries

Country	Population 2013	Population in densely populated areas 2012 (%)	Average household size 2012	GDP per capita 2013	Housing system	Welfare regime
Denmark	5,602,628	33.7	1.9	45,100	Unitary	Social democratic
Germany	82,020,578	35.3	2.0	34,200	Unitary	Conservative/corporatist
Hungary	9,908,798	29.6	2.6	10,200	'Super' home ownership	Transition
Ireland	4,591,087	35.1	2.7	38,000	Dual	Liberal/ conservative
Netherlands	16,779,575	47.3	2.2	38,300	Unitary	Social democratic/ corporatist
Norway	5,051,275	53.3	2.1	77,400	Unitary	Social democratic
Romania	20,020,074	33.7	2.9	7,200	'Super' home ownership	Transition
Spain	46,727,890	50.4	2.6	22,500	Familialism	Mediterranean/ Southern European
Sweden	9,555,893	20.8	2.1	45,500	Unitary	Social democratic
Switzerland	8,039,060	26.2	2.3	63,800	Unitary	Liberal
UK	63,896,071	57.0	2.3	31,500	Dual	Liberal
EU28	50,661,827	42.1	2.4	25,500	N.A.	N.A.

Sources: Eurostat; Welfare regime and Housing system – Fitzpatrick and Stephens, 2014.

Notes: Dates refer to latest year for which data are available for most countries; GDP – current market prices (euro).

Ireland and Spain were two of the European countries most affected by the GFC. In the Irish chapter, Winston outlines the highly significant impacts of the bust and subsequent recovery on the sustainability of urban communities. Some significant problems include: housing affordability; a very low proportion of the housing stock in multi-family dwellings; relatively poor public transport; fuel poverty among certain groups; and a high dependence on fossil fuels for energy. As in the Irish case, the Spanish chapter reflects on how sustainable communities have been challenged by recent real estate expansion where housing was employed as a driver of economic growth and employment. Pareja-Eastaway and Sánchez-Martínez examine the negative social consequences of the Spanish crisis, including very significant levels of unemployment and debt. In addition, attention has shifted away from environmental matters and holistic approaches to sustainability and economic concerns are prioritised.

Sweden has introduced a range of policies targeting environmental sustainability over a long period of time, including those focusing on reducing energy consumption in housing. This has resulted in a relatively high-quality, energy efficient housing stock there. However, Werner expresses some concern about increasing inequalities in a number of areas in Sweden, including access to housing, housing quality, dwelling space and security of tenure. Expanding the supply of affordable housing is highlighted as a key issue. In the Norwegian chapter, Støa outlines difficulties relating to pathways to a low carbon society in another affluent country. Rising housing consumption patterns are highlighted as a significant problem in Norway. The chapter emphasises the implications for residential development and explores the role of the quality of sustainable housing and communities as a key driver of progress towards more sustainable societies. Larsen and colleagues present a paradox as Denmark is considered a front-runner when it comes to sustainable communities and housing, yet the Danes are among those European consumers with the greatest ecological footprint. The chapter examines the role of public regulation versus civil society (local organisations) as a driver of more sustainable communities.

In the chapter on the Netherlands, Kleinhans notes that the creation of sustainable housing was an important element of Dutch national urban regeneration policy for many years. Housing associations were a key driver of this, including in their work on increasing energy efficiency in the social housing sector. Financial austerity post-GFC has decreased the investment capacities of housing associations, limiting their ability to contribute to sustainable housing and communities. In the chapter on the UK, Bramley argues that housing conditions and policies are diverging across its constituent countries and regions. This is occurring against a backdrop of significant structural economic weaknesses and high levels of inequality. The key housing issues are undersupply, affordability and declining home ownership. A key challenge is to obtain public support for more sustainable development in this decentralised planning system. In the interim, historic improvements in housing conditions are being reversed due to the combined effects of increased demand, the GFC and undersupply.

Both the Romanian and Hungarian chapters discuss socialist housing policies and relevant changes since 1989. In Romania, Turcu outlines how state-provided mass housing tended to consist of poor-quality apartments built close to industry. Attempts were made to consolidate urban settlements by increasing densities, building new urban housing districts at higher densities with no green space. Privatisation in the early 1990s involved sitting tenants buying state-owned housing at massively discounted prices, resulting in a significant rise in home ownership and 'residualisation' of public housing. Current housing policy aggressively promotes home ownership, with scant attention given to private renting, and the social rented sector is almost non-existent. In the Hungarian chapter, Tosics argues that the collapse of the socialist model brought contradictory changes regarding sustainable housing. Although housing standards and quality of life increased for many, social and territorial differentiation rose too. Housing policy has not been the focus of policy-making over this period, becoming more 'residual', with an unstated assumption that the market would solve the problem. A lack of political will to regulate the market in the public interest, including environmental and social sustainability, has led to a number of problems which are highlighted in the chapter.

The Swiss housing market has been largely unaffected by the housing and economic crisis compared with some other European countries over the last ten years. A key challenge there, according to Hugentobler, is the limited supply of affordable housing. Demand for housing has risen due to economic growth, immigration and the attractions of urban living. There are also problems regarding the renovation and renewal of the older housing stock. However, some local policies in larger cities are providing the context for sustainable housing projects, offering high-quality urban living run by non-profit housing cooperatives, municipalities and other pioneering organisations.

In Germany the divide between East and West is still most evident when assessing sustainable communities. Deilmann and Effenberger highlight gaps between various strategies in Germany and actual results for land take and resource consumption. They identify consumers as the principal drivers of ever-increasing resource consumption. However, the authors argue that key future challenges for the sustainability of communities will arise less from ecological issues and more from social and economic issues due to demographic change, migration and socio-economic divisions.

As mentioned above, in the final section of each country chapter the authors discuss an issue of particular relevance or importance in their country. In the chapter on Ireland, Winston examines measures to address energy poverty, including an assessment of relevant building regulations and retrofitting programmes. In the Spanish chapter, Pareja-Eastaway and Sánchez-Martínez examine the impacts of massive second-home developments during the period of economic expansion (2000–2007). In the chapter on Sweden, Werner explores the social impacts of the 'right to buy' policy available for municipal housing tenants, focusing on four case studies in Stockholm. In the Norwegian chapter, Støa examines lessons learned from a, to-date, unsuccessful attempt to establish a new sustainable urban neighbourhood in Trondheim. The Danish authors compare four approaches to sustainable housing in Denmark including: a municipality's initiative to transform a town into a zero-carbon town; a grassroots community cooperative project; social housing retrofitting; and a voluntary sustainable neighbourhood certification scheme. Kleinhans examines the potential of community enterprises in the Netherlands to contribute to sustainable communities. In the chapter on the UK, Bramley focuses on energy efficiency and fuel poverty, but also on social sustainability, urban form and social mix. In her chapter on Romania, Turcu analyses a programme to energy retrofit apartment buildings. Tosics assesses plans to combat social segregation in Hungary. In the chapter on Switzerland, Hugentobler focuses on the role of non-profit housing cooperatives in Zurich. In their chapter, Deilmann and Effenberger examine the issues of housing vacancies and shortages in Germany. The final chapter revisits the key themes raised in this Introduction and presents some conclusions regarding sustainable housing definitions, indicators, policies and practices in the countries examined.

References

Baker, S. (2015) *Sustainable Development*. Second edition. London: Routledge.

Beck, U. (1992) *Risk Society, Towards a New Modernity*. London: Sage.

Betsill, M. and Bulkeley, H. (2007) 'Looking back and thinking ahead: a decade of cities and climate change research', *Local Environment: The International Journal of Justice and Sustainability*, 12(5):447–456.

Boelhouwer, P. and Priemus, H. (2014) 'Demise of the Dutch social housing tradition: impact of budget cuts and political changes', *Journal of Housing and the Built Environment*, 29:221–235.

Bulkeley, H. (2010) 'Cities and the governing of climate change', *Annual Review of Environment and Resources*, 35:229–253.

Bulkeley, H., Castan Broto, V., Hodson, M. and Marvin, S. (2013) *Cities and Low Carbon Transitions*. London: Routledge.

Cisneros, H., Dyer-Chamberlain, M. and Hickie, J. (eds) (2012) *Independent for Life: Homes and Neighborhoods for an Aging America*. Austin: University of Texas Press.

Commission of the European Communities (CEC) (2015) *Legal Enforcement: Statistics on Environmental Infringements*. Available at http://ec.europa.eu/environment/legal/law/statistics.htm, accessed 6 July 2016.

Commission of the European Communities (CEC) (2013) *A 2030 Framework for Climate and Energy Policies* (COM (2013) 0169). Brussels: European Commission.

Edwards, B. and Turrent, D. (eds) (2000) *Sustainable Housing: Principles and Practice*. London: E. & F. N. Spon.

Elsinga, M. and Lind, H. (2012) 'The effect of EU legislation on rental systems in Sweden and the Netherlands', *Housing Studies*, 28(7):960–970.

European Environment Agency (2011) *Greenhouse Gas Emission Trends and Projections in Europe 2011*. Available at www.ab.gov.tr/files/ardb/evt/1_avrupa_birligi/1_6_raporlar/1_3_diger/environment/ghg_emmissions_trends_and_projections.pdf, accessed 6 July 2016.

Fitzpatrick, S. and Stephens, M. (2014) 'Welfare regimes, social values and homelessness: comparing responses to marginalised groups in six European countries', *Housing Studies*, 29(2):215–234.

Gruis, V. and Elsinga, M. (2014) 'Tensions between social housing and EU market regulations', *European State Aid Law Quarterly*, 3(14):463–469.

Housing Europe (2015) *The State of Housing in the EU 2015*. Brussels: Housing Europe.

Huby, M. (1998) *Social Policy and the Environment*. Buckingham: Open University Press.

Huby, M. (2002) 'The sustainable use of resources', in T. Fitzpatrick and M. Cahill (eds) *Environment and Welfare: Towards a Green Social Policy*, pp. 117–137. New York: Palgrave.

Jordan, A. and Adelle, C. (eds) (2013) *Environmental Policy in the EU*. Third edition. London: Earthscan from Routledge.

Jordan, A. and Tosun, J. (2013) 'Policy implementation', in A. Jordan and C. Adelle (eds) *Environmental Policy in the EU*, pp. 247–266. Third edition. London: Earthscan from Routledge.

Kallis, G. (2011) 'In defence of degrowth', *Ecological Economics*, 70(5):873–880.

Kemeny, J. (1995) *From Public Housing to the Social Market: Rental Policy Strategies in Comparative Perspective*. London: Routledge.

Kemeny, J. Kersloot, J. and Thalmann, P. (2005) 'Non-profit housing influencing, leading, and dominating the unitary rental market: three case studies', *Housing Studies*, 20(6):855–872.

Knill, C., Heichel, S. and Arndt, D. (2012) 'Really a front-runner, really a straggler? Of environmental leaders and laggards in the European Union and beyond: a quantitative policy perspective', *Energy Policy*, 48:36–45.

Lock, D. (2000) 'Housing and transport', in B. Edwards and D. Turrent (eds) *Sustainable Housing: Principles and Practices*, pp. 36–42. London: Taylor and Francis.

Meadowcroft, J. (2007) 'National sustainable development strategies: features, challenges and reflexivity', *Environmental Policy and Planning*, 9(3–4):193–212.

Norman J., Maclean, H. and Kennedy, C. (2006) 'Comparing high and low residential density: life-cycle analysis of energy use and greenhouse gas emissions', *Journal of Urban Planning and Development*, 132(1):10–22.

OECD (2009) *Strategies for Sustainable Development*. Available at www.oecd.org/document/40/0,3343,en_2649_34421_2670312_1_1_1_1,00.html, accessed 6 July 2016.

Office of the Deputy Prime Minister (ODPM) (2005) *Sustainable Communities: People, Places, and Prosperity* (Bristol Accord). London: ODPM.

Pareja-Eastaway, M. and Støa, E. (2004) 'Dimensions of housing and urban sustainability', *Journal of Housing and the Built Environment*, 19:1–5.

Rudlin, D. and Falk, N. (2009) *Sustainable Urban Neighbourhood: Building the 21st Century Home*. Second edition. Oxford: Elsevier Architectural Press.

Schneider, F., Kallis, G. and Martinez-Alier, J. (2010) 'Crisis or opportunity? Economic degrowth for social equity and ecological sustainability', *Journal of Cleaner Production*, 18:511–518.

Späth, H. and Rohracher, P. (2013) 'The eco-cities Freiburg and Graz: the social dynamics of pioneering urban energy and climate governance', in H. Bulkeley, V. Castan Broto, M. Hodson and S. Marvin (eds) *Cities and Low Carbon Transitions*, pp. 88–106. London: Routledge.

Stead, N. (2000) 'Unsustainable settlements', in H. Barton (ed.) *Sustainable Communities: The Potential for Eco-neighbourhoods*, pp. 29–39. Second edition. London: Earthscan.

Swanson, D., Pinter, L., Bregha, F., Volkery, A. and Jacob, K. (2004) *National Strategies for Sustainable Development: Challenges, Approaches and Innovations in Strategic and Co-ordinated Action: A 19 Country Study*. Winnipeg: International Institute for Sustainable Development.

Turcu, C. (2009) 'In the quest of sustainable communities: a theoretical framework to assess the impact of urban regeneration', in S. Tsenkova (ed.) *Planning Sustainable*

Communities: Diversity of Approaches and Implementational Challenges, pp. 37–66. Calgary: University of Calgary.

Wheeler, S. (2004) *Planning for Sustainability: Creating Livable, Equitable, and Ecological Communities*. Abingdon: Routledge.

Winston, N. (2007) 'From boom to bust? An assessment of the impact of sustainable development policies on housing in the Republic of Ireland', *Local Environment*, 12(1):57–71.

Winston, N. (2010) 'Urban regeneration and housing: challenges for the sustainable development of Dublin', *Sustainable Development*, 18:319–330.

Winston, N. (2014) 'Sustainable communities? A comparative perspective on urban housing in the European Union', *European Planning Studies*, 22(7):1384–1406.

Winston, N. and Pareja-Eastaway, M. (2008) 'Sustainable housing in the urban context: international sustainable development indicator sets and housing', *Social Indicators Research*, 87:211–221.

World Commission on Environment and Development (WCED) (1987) *Our Common Future*. Oxford: Oxford University Press.

Worpole, K. (2003) 'The social dynamic', in P. Neal (ed.) *Urban Villages and the Making of Communities*, pp. 119–131. London: Spon Press.

Worpole, K. and Knox, K. (2007) *The Social Value of Public Spaces*. York: Joseph Rowntree Foundation.

2 Republic of Ireland

Nessa Winston

Introduction

The Republic of Ireland (hereafter referred to as Ireland) is the smallest of the countries examined in this book with a population of 4.6 million (CSO, 2012b: 13). Sixty-one per cent live in urban areas, locations with population of 1,500 or more, and approximately one-third (35 per cent) lives in densely populated areas, somewhat below the EU average of 42 per cent (see Table 1.1). Ireland's recent macroeconomic history has been referred to as a 'rollercoaster' (Whelan and Maitre, 2014). The 'Celtic Tiger' period of very substantial economic growth from the mid-1990s–2007 was preceded by two decades of economic stagnation and under-performance (Kennedy *et al.*, 1988). Despite this, significant employment growth was achieved during the 'Celtic Tiger' era and the population increased by 17 per cent and households by 14 per cent (CSO, 2007).

However, since 2007, a range of factors has had substantial social and economic impacts on individuals and households in Ireland. These include the bursting of a national credit 'bubble', the effects of the Global Financial Crisis (GFC) and government responses to the inability of Irish banks to raise finance, in particular the guarantee of all bank deposits and senior debt, and the full or part nationalisation of all but one of the Irish-based banks and building societies. The cost of this 'bailout' and an increasingly negative exchequer balance undermined Ireland's sovereign creditworthiness, despite a very severe programme of fiscal austerity. In November 2010, the Irish government signed up for an emergency loan from the International Monetary Fund (IMF) and the EU to finance public spending and the bank recapitalisation programme. GDP declined by 13 per cent between 2008 and 2011. Unemployment rose from 4 to 14 per cent in spite of quite substantial net emigration (Callan *et al.*, 2013; CSO, 2012a). At the time of writing, the country is emerging from the recession and, by comparison with its European neighbours, it is relatively wealthy, as indicated by its GDP per capita (Table 1.1). Unemployment was 8.8 per cent in February 2016, down from a peak of 15.2 per cent in January 2012, although youth unemployment remains relatively high at 20.1 per cent (CSO, 2016). The latest IMF report on Ireland notes the robust growth but cautions about the high levels of public and private debt and the importance of continued reductions in same (IMF, 2015).

Ireland's social security system provided considerable protection to many during the GFC. This system is frequently classified as a liberal or conservative welfare regime (Arts and Gelissen, 2010; Dukelow, 2011), which is not surprising given that it emerged during the period of British rule and that British influence remained strong post-independence (Daly and Yeates, 2003). However, McCashin (2012) has highlighted problems with this classification, and Hick (2014) points out that the range and levels of social security payments expanded significantly in real terms during the Celtic Tiger era. During the crisis, there were cuts in some social security payments, especially those for younger claimants of Jobseeker's Allowance, but also to the universal child benefit payment. However, many core welfare payments were protected (Nolan *et al.*, 2014) and Hick (2014: 406) concludes that 'many, if not all, of the gains of the pre-crisis period have been preserved, thus far at least'. Similarly Callan *et al.* (2013) find that those towards the bottom of the social class scale, already quite dependent on social transfers prior to the recession, improved their relative position on the scale considerably. During the boom, welfare expansion was accompanied by tax cuts. However, the GFC resulted in an increase in the range and extent of taxes, along with a number of cuts to public sector salaries. Economic vulnerability increased and became more widely distributed across the social classes (Whelan and Maitre, 2014: 482).

Ireland has, and actively promotes, an image of being the 'Emerald Isle'. Its score on the Environmental Protection Index is slightly above the average for the EU28 (Table 1.1). Prior to EU membership in 1973, environmental regulations were limited, and since accession its track record regarding EU environmental law has been criticised (e.g. Flynn, 2007). A recent European Commission (2015) report noted that Ireland was likely to reduce emissions by just 3 per cent by 2020 when the target was 20 per cent. Davies (2013: 65) suggests that Irish attitudes to natural resources are utilitarian and that conservation and regulation are a 'preoccupation of the privileged, often urban, elites and an interference with property rights'. However, one survey found improving attitudes and behaviours among a representative sample of Irish adults (Motherway *et al.*, 2003), while another study revealed increased acceptance of the benefits of certain relevant environmental directives among some farmers (see Buckley, 2012). The latter is significant given that one of the two most problematic areas for carbon emissions is agriculture, the other being transport.

The housing system in Ireland

Housing policy and legislation are the responsibility of the national Housing Ministry while implementation of housing programmes is the remit of local authorities and approved housing bodies (AHBs). Current national housing policy aims 'to enable all households access good quality housing appropriate to household circumstances and in their particular community of choice' (DECLG, 2011: n.p.). This represents a departure from a long-standing policy of promoting home ownership which has resulted in owner occupation being the dominant tenure (Table 2.1). The tenure expanded for most of the twentieth century due to

Table 2.1 Housing tenure, Republic of Ireland, selected years (%)

Tenure	1946	1971	1991	2006	2011
Owner occupied	52.6	70.8	80.0	77.2	70.8
Private rented	26.1	10.9	8.1	11.0	18.8
Local authority rented	16.5	15.9	9.8	7.2	7.9
Rented – voluntary body	–	–	–	3.5	0.9
Other	4.7	2.4	2.1	1.5	1.6
Total	100	100	100	100	100
Number of units (,000)	666	726	1,006	1,462	1,649

Source: Census of population, various years.

a range of substantial supports from government, introduced prior to independence from Britain in 1922. The supports were expanded to such an extent that Norris and Winston (2011: 18) argue they 'amounted effectively to a "socialised" system of home ownership'. They included a large number of very attractive local authority tenant purchase schemes and loans. Fahey (1999) estimated that approximately one-third of local authority stock was sold to tenants. Following a fiscal crisis in the 1980s, many of these public supports were withdrawn. During the Celtic Tiger era, a combination of social and economic factors (including increased employment and affluence) led to a very significant rise in the demand for housing and unprecedented house price inflation. This, combined with the withdrawal of home ownership subsidies, resulted in a decline in home ownership from its peak of 80 per cent in 1991 to 71 per cent in 2011. A housing policy statement in the latter year announced a policy of 'equity across tenures', ending the explicit promotion of home ownership in government policy.

Local authorities have been the main providers of social housing in Ireland. Local authority tenants pay differential rents based on household income – on average about 15 per cent of this income (DECLG, 2014: 8). Table 2.1 reveals a decline in local authority housing from 17 per cent in 1946 to 8 per cent in 2011. Following a period of expansion from the early 1930s to the 1960s, the stock was considerably reduced due to public spending cuts and the long-standing policy of sales to tenants mentioned above. Tenant purchase dates back to the late nineteenth century, pre-independence, when it was introduced to appease nationalist tenant farmers in rural Ireland (Fahey, 2002). However, it was also due to strong mobilisation by urban tenants for more favourable rents and for purchase schemes (Hayden, 2014a). As a result, the sector became increasingly unattractive to the state (Hayden, 2014a: 117).

The voluntary sector, approximately 500 AHBs, plays a small but increasingly important role in social housing provision. Funding for this sector was very limited until the 1980s but schemes introduced in 1984 and 1991 provided a boost to their capacity. The GFC resulted in a sharp rise in the number of households on the waiting list for social housing between 2005 and 2011 (DECLG, various years). At the same time, there were severe cuts to exchequer funding for social housing (DECLG, 2014: vii). The 2011 housing policy statement

(DECLG, 2011: 3) stated that financial circumstances 'rule out a return to very large capital-funded construction programmes by local authorities' and put AHBs 'at the heart of the Government's vision for housing provision'. In 2013, 89,872 households qualified for social housing support, with just over one-third in the Dublin area (DECLG, 2014: 6–7). By then, the private rented sector was the main way in which social housing was provided. The 2014 Social Housing Strategy set a target of 35,000 new social housing units, with the 'two primary delivery channels' being local authorities and AHBs (DECLG, 2014: viii), but it reiterated an 'enhanced role' for AHBs, following a number of changes to the sector, including the introduction of multi-annual housing expenditure programmes, prioritisation of funding to incentivise scale, and enhanced regulation (DECLG, 2014: ix).

The private rented sector, the majority tenure in the pre-independence period, was relatively neglected in Ireland for a long period, to such an extent that it was termed 'the forgotten sector' (O'Brien and Dillon, 1982). The sector was relatively unregulated and supports for it were relatively limited compared with those for home ownership (FitzGerald and Winston, 2011). Stock declined from 26 per cent in 1946 to 7 per cent in 1991. However, the sector has been expanding since the 1990s for a variety of reasons, including the introduction of a range of tax incentives and improved, though still far from perfect, regulation. The increase is also due to more limited access to both home ownership and social housing. A number of government housing supports are provided to tenants in the private rented sector and it is estimated that more than half of all rents received by private landlords stem from these schemes (DECLG, 2014: x).

Ireland has been categorised in the dualist rental model by Kemeny (1995). This model is one in which there are significant differences between the social and private rented sectors, so it contrasts with the integrated/unitary model, where there is little difference between the two. However, Norris (2014a) argues that the dual model in Ireland has been replaced by an 'embryonic' version of the unitary model due to the withdrawal of public subsidies for home ownership in the 1980s and the abolition of social housing subsidies during the GFC. She argues that there is increasing tenure neutrality across the rented sector in relation to: housing subsidies; secure occupancy rights; and minimum standards regulations.

Sustainable communities and housing in Ireland

With the introduction of the Planning and Development Act 2000, sustainable development became enshrined in Irish legislation. However, Grist (2012: 8) notes that the term is not defined as the relevant minister argued that the 'concept is dynamic and all-embracing and will evolve over time, so a legal definition would tend to restrict and stifle it'. Sustainable communities were defined in a 2007 Housing Ministry policy statement as:

> places where people want to live and work, now and in the future . . . meet the diverse needs of the existing and future residents, are sensitive to their

environment, and contribute to a high quality of life. They are safe and inclusive, well-planned, built and run, offer equality of opportunity and good services to all.

(DEHLG, 2007b: 7)

This reflects the 'Bristol Accord' (ODPM, 2005), and the statement specifically refers to it. It adds: 'Sustainable communities have a high quality natural and built environment, with a dynamic and innovative economy, good transport, supportive community and voluntary services, and are environmentally sound' (DEHLG, 2007b: 21). The document included the need for 'sufficient homes of the right type, in the right locations, integrated with other infrastructure requirements', to 'improve the environmental sustainability of new housing through voluntary codes and building regulations' and investment in high-quality mixed tenure estates and social housing renewal, maintenance and management (DEHLG, 2007b: 11–12). Drawing on a 2004 National Economic and Social Council (NESC) report, it defined sustainable neighbourhoods as:

> areas where an efficient use of land, high quality urban design and effective integration in the provision of physical and social infrastructure such as public transport, schools, amenities and other facilities combine to create places people want to live in. Additional features of sustainable neighbourhoods include: compact, energy efficient and high quality urban development; accessibility via public transport networks and also meeting the needs of the pedestrian and cyclist; and provision of a good range of amenities and services within easy and safe walking distance of homes.

(DEHLG, 2007b: 27)

The 2014 Social Housing Strategy, referring back to the NESC vision quoted above, stated that the goals of housing policy should include sustainability 'economically, social and environmentally' (DECLG, 2014: 17). It added tenure mix with a wide range of accommodation types, noted that design priorities should be 'socially and environmentally appropriate, architecturally appropriate, safe, secure and healthy; affordable; durable; and accessible and adaptable' (DECLG, 2014: 5). There were no specific references to the natural environment, and it is not clear what the term 'environment' means.

A white paper on energy states that energy efficiency will be central to Ireland's transition to a low carbon economy (DCENR, 2015a). Other relevant goals include: ensuring that everyone can afford adequate energy and a warm home (DCENR, 2007, 2011, 2015b); that all new housing will be carbon neutral; that energy efficiency in older homes will be significantly improved via retrofitting schemes (DCENR, 2009); and the implementation of cost-effective approaches to combat fuel poverty (DCENR, 2015a). Finally, in line with Directive 2010/31// EU on the energy performance of buildings (recast), all new buildings will be nearly zero energy by 2020 (DECLG, 2012). This will mean they should have a BER of A3 or higher.

In general, national policy and guidelines for local authorities now incorporate many of the characteristics of sustainable housing (DEHLG, 2007c, 2009a, 2009b), although the 2007 ones were somewhat more limited (Winston, 2008). The question arises about the extent to which Irish housing and communities may be considered sustainable in practice. The next sub-sections examine this with particular focus on: land use planning, settlement and commuting patterns; housing quality, energy and fuel poverty; and housing affordability.

Land use planning, settlement and commuting patterns

Some serious shortcomings in land use planning were the subject of recent tribunals of inquiry in Ireland (Mahon *et al.*, 2012; Grist, 2012). These included corruption in the planning and development areas which gave rise to over-zoning of land for housing, at 4.5 times actual need (Grist, 2012). Housing stock per capita was the lowest in the EU in 2000 (Norris and Winston, 2003), but stock increased by 72 per cent between 1991 and 2011 (CSO, 2011). Key drivers of this were the ready availability of credit, a range of property-based tax incentive schemes and house price inflation. Some of this property was inappropriately located, distant from essential public services, and/or built on flood plains, resulting in serious problems with, or risk of, flooding. By 2008, there were a large number of unfinished developments and 'ghost' estates all over the country. Unfinished housing estates are defined as a housing estate of two or more housing units where development and services had not been completed and estates completed from 2007 onwards where 10 per cent or more of units were vacant. A ghost estate is a more extreme version, defined by Kitchen *et al.* (2014) as ten or more housing units where 50 per cent or more units are either vacant or under construction. In October 2011 there were 2,876 unfinished estates (122,048 units), 777 of which were ghost estates (Kitchin *et al.*, 2014). Problems associated with them include health and safety issues, antisocial behaviour and governance issues (Kitchen *et al.*, 2014; Mahon and O'Cinneide, 2010).

Even large urban areas, such as the Greater Dublin area, have grown via accelerated urban sprawl in the more distant commuter belts (Williams *et al.*, 2007; Williams and Shiels, 2000). This may have facilitated the construction of larger homes – residential dwellings are larger than the EU average (see Table 2.3), and the built form is quite distinctive by European standards. Approximately 9 per cent of dwellings are classified as multi-family housing, the lowest in the EU, where the average is 42 per cent (Dol and Haffner, 2010). Census figures reveal considerable changes in the type of housing constructed over the past decade. Detached dwellings declined from 54 per cent of stock in 1990 to 43 per cent in 2011. Semi-detached houses increased from 19 to 28 per cent of stock, and apartments from 4 to 11 per cent. However, terraced houses declined from 23 to 17 per cent.

A very substantial rise in commuting by car has occurred over time so that it is now by far the dominant means of transportation (Table 2.2). These figures mask variations by area but even within urban areas there are often problems of access

Table 2.2 Means of travel to work, education (persons aged five years and over), Ireland (%)

	1981	1986	1991	1996	2002	2006	2011	Change
On foot	18.1	16.7	15.2	14.3	13.4	12.8	11.4	−6.7
Bicycle	5.6	7.5	6.1	4.5	2.5	2.3	2.7	−2.9
Bus	13.1	11.5	10.5	9.5	7.8	7.1	6.1	−7.0
Train	1.7	2.0	2.4	2.2	2.4	3.4	3.5	1.8
Motorcycle	2.1	1.9	1.5	1.2	1.3	0.8	0.6	−1.5
Car	59.4	60.5	64.3	68.5	72.6	73.6	75.8	16.4

Source: Census of population, various years.

to good-quality public transportation. For example, Wickham (2006) describes Dublin as a 'car dependent' city, which he argues exacerbates social exclusion as people have to make sacrifices to run a car and the lack of a car can result in people being 'trapped' in their local community.

Housing quality, energy, and poverty

Housing quality can have a significant impact on the energy efficiency of dwellings and carbon emissions but also on residents' health and energy poverty risk. Irish housing quality improved significantly over a number of decades and is now relatively good by European standards (Table 1.2; see also Norris and Domanski, 2009). This is due to a combination of factors, including improved building regulations, the extent of recent construction and rising household incomes. However, there are some energy-related problems, including a high dependence on imported fossil fuels, relatively high energy consumption levels and energy poverty. The residential sector accounts for 28 per cent of all energy-related CO_2 emissions (O'Meara, 2015). Import dependence increased significantly from the mid-1990s such that 89 per cent of total energy was imported in 2013 (SEAI, 2014: 47). Fossil fuels accounted for 91 per cent of all energy used in 2013 (SEAI, 2014: 7). In the past, the dominant forms of home heating were open fires and back-boiler systems fuelled by coal and/or peat. Housing built since the 1990s has tended to have oil- or gas-fired central heating or, less frequently, electric storage heating, and since the late 1980s there have been considerable advances in converting back-boilers in older housing to oil or gas. Oil is now the dominant residential fuel (33 per cent), followed by electricity (25 per cent) and natural gas (22 per cent). Irish housing is much more dependent on oil and gas than other countries (ODYSEE, n.d.). There has been a substantial decline in coal and peat usage, but they are still relatively prominent – 16 per cent of final energy use compared with 0 and 2 per cent in Germany and the UK, respectively (ODYSSEE, n.d.). The role of renewables is still very limited, the third lowest in the EU for direct use (2 per cent) and fifth lowest for direct and indirect use (4 per cent) (ADEME, 2015: 87–88). On a positive note, there has been some improvement in the use of solar water heaters since 2000 (ADEME, 2015: 89).

Table 2.3 Residential energy indicators, Ireland, 2000–2013, Ireland versus EU and UK, 2013

	Unit consumption per dwelling with climatic corrections (toe/dw)	Unit consumption per dwelling for space heating with climatic corrections (toe/dw)	Unit consumption per m² for space heating with climatic corrections (koe/m²)	Floor area of dwellings (average, m²)
2000	2.02	1.41	13.102	107.455
2001	2.08	1.43	13.149	109.063
2002	2.12	1.49	13.582	110.018
2003	2.09	1.46	13.156	110.769
2004	2.1	1.45	12.921	112.075
2005	2.13	1.48	12.998	113.497
2006	2.11	1.45	12.602	114.947
2007	2.07	1.43	12.231	116.586
2008	1.96	1.32	11.193	117.798
2009	1.87	1.27	10.695	118.562
2010	1.74	1.15	9.688	118.971
2011	1.71	1.14	9.526	119.341
2012	1.56	1.02	8.507	119.61
2013	1.56	1.03	8.552	119.865
EU 2013	1.42	0.97	10.802	89.748
UK 2013	1.45	0.96*	10.41*	95.287

Source: Collated by the author from Odyssee (n.d.) database.

Note: * Figures refer to 2012.

There is some evidence to suggest that Irish energy consumption levels are relatively high. Climate corrected figures for total energy use/dwelling and energy use per dwelling for space heating are higher than the EU and UK figures (Table 2.3), although they have been declining since 2005/2006. During the Celtic Tiger era, household incomes rose considerably, which may have increased the number and use of household appliances, and the extent to which people heated their homes. In addition, average dwelling size in Ireland is larger than in the EU, which may be due, in part, to a slightly larger household size in Ireland (Table 1.1). Ireland fares better on energy use per square metre, which has declined since 2005, although it increased slightly in 2013. Energy consumption per dwelling has declined in the vast majority of EU member states since 2000, which is linked to energy efficiency improvements, higher energy prices since 2004 (+64 per cent) and the recession, but Ireland is among the countries that have experienced a 'very strong reduction' (above 4 per cent) (ADEME, 2015: 23–24). The decline has been much more significant since 2008, which suggests that changes in behaviour due to higher prices and lower incomes are a more important driver of the decline than other factors. Similarly, a significant decline in space heating per square metre during this period has been attributed to cost and ability to pay (ADEME, 2015: 30). However, over a longer period (1990–2012), reduced solid fuel use has had a substantial impact on unit consumption (ADEME, 2015: 33). Furthermore, Ireland's efficiency ranking improves considerably when the use of condensing boilers and heat pumps is taken into account – it is among the most efficient on this indicator (ADEME, 2015: 34). This may be linked to the introduction of minimum efficiency standards for condensing boilers: Ireland is one of very few countries to have implemented this measure (ADEME, 2015: 58).

In Ireland, energy poverty is defined as an inability to heat one's home to an adequate degree which is linked to income, energy costs and the energy efficiency of the home (DCENR, 2015a: 32). The Government Strategy on Affordable Energy (DCENR, 2011) employed the expenditure method (household has to spend more than 10 per cent of income on energy) to assess the problem. This revealed that over 20 per cent of households experienced the problem in 2009 (DCENR, 2015a: 32). This will have risen during the GFC. SILC data for 2014 reveal that 16 per cent went without heating in the previous year, up from 6 per cent in 2007, and a further 9 per cent were unable to afford to keep their homes adequately warm (CSO, 2015). Cultural factors may partly explain the levels of reporting on these subjective indicators (Bouzarovski, 2011). The most recent data for Ireland use the objective method, which estimates energy poverty by modelling what a typical household must spend to keep the home heated to World Health Organisation norms and compares this to household incomes. It found 28 per cent of households in fuel poverty, with a higher prevalence among social housing tenants, and in homes heated by oil and solid fuel (DCENR, 2015a: 34). High and fluctuating prices for those fuel sources are important factors in the problem (DCENR, 2015a: 26).

Housing affordability

Despite very substantial levels of new housing output during the Celtic Tiger era, there was significant inflation in house prices and private rents, and increased demand for social housing. Mortgage lending increased much more in Ireland than elsewhere (Doyle, 2009; Kelly, 2009). This included increases in both the size and variety of mortgages (including interest-only and 100 per cent mortgages) and a decline in lending standards among mainstream lenders (Norris and Coates, 2014). During the recession house prices and private rents declined significantly while affordability improved (see NESC (2014a) for a detailed discussion of affordability). A number of indicators suggest that, overall, housing is relatively affordable by comparable international standards (Radi Rogerova *et al.*, 2014; Cox and Pavletch, 2013; Demographia, 2016). However, the economic recovery has led to both house price and private rent inflation, especially in Dublin (but also in other cities). Combined with a shortage of social housing, this has resulted in affordability problems for certain groups. For example, while the Irish housing cost overburden rate (6 per cent) was the lowest of the countries examined in this volume, the rate was significantly higher for those in the lowest income quintile, single person households under 65 years (both 19 per cent), private renters (18 per cent) and single parents (12 per cent) (see Table 2.4).

Table 2.4 Groups with above-average housing cost overburden rate (%), Ireland, 2013

Category	Housing cost overburden rate (%)
Total sample	6.0
First income quintile	19.2
Single adult under 65 years	18.7
Private renting	17.8
Single person with dependent children	11.9

Source: Compiled by the author from Eurostat SILC.

Note: Housing cost overburden rate = percentage of population living in a household where total housing costs (net of housing allowances) > 40 per cent of total disposable household income (net of housing allowances).

Another indicator of affordability is the extent of rent or mortgage arrears. In 2013 such arrears were more than twice the EU28 average (see Table 1.2). Despite a high level of mortgage arrears for principal dwelling homes (PDH) during the GFC, repossessions were relatively low. This was primarily due to government agreements with mortgage lenders via the banking sector recapitalisation programme but also to increased take-up of mortgage interest supplement, a means-tested support for unemployed homeowners. There has been some improvement in arrears over recent years. In June 2013, 142,892 PDH accounts were in arrears and 12.7 per cent of loan accounts were in arrears over 90 days. By December 2015, these figures had declined to 88,000 and 8.3 per cent (Central Bank of Ireland, 2015). However, mortgage arrears are still a sizeable problem, indicating that home ownership may not be a sustainable option for many households who purchased during the boom years.

Towards more sustainable communities and housing: barriers, drivers and challenges

There are a number of challenges to the provision of more sustainable communities in Ireland. These include fixing the mistakes of the past, where possible, and increasing the supply of good-quality, affordable housing in appropriate locations, accessible to public transport and other key services. Some progress has been made on a number of these issues, but in other areas considerable barriers remain which impede the transition to more sustainable housing. With regard to unfinished housing developments, a Government Action Programme has reduced their number by 75 per cent over a five-year period and vacancy rates by 31 per cent between 2014 and 2015 (Housing Agency and DECLG, 2015). Demand for those located close to centres of employment is likely to result in the completion and occupation of many more but this is much less likely to be the case for those in more peripheral regions. With regard to future supply, changes to planning law between 2010 and 2012 led Grist (2012: 179) to argue that there will be 'better realignment of the quantum of zoned land with population and housing targets'. With regard to new housing, there is now a very good set of sustainable community design guidelines, capturing many of the key dimensions of sustainable housing. In addition, in line with the EU Floods Directive, robust guidelines for flood risk management have been introduced (DEHLG and OPW, 2009). Not much housing has been built in the interim. It is imperative that all new housing construction is in keeping with the sustainable community and flood risk management guidelines as well as local climate change adaptation plans.

With a rising population and decreasing household size, increasing the supply of higher-density, high-quality urban housing, designed for different types of households, including families, is most important. Data on built form reflect negative perceptions and/or experiences of some apartments to date, including their suitability for families and long-term residence (Howley *et al.*, 2009; Norris and Winston, 2009). Until 2007, space standards and other aspects of design were relatively modest compared with a number of other European countries. Improved national guidelines were introduced in 2007 and made mandatory in 2015. These increased minimum floor space by approximately 15 per cent, but very little has been built under them to date (DEHLG, 2007a). Additional requirements were made in 2015, including the introduction of studio units in managed, built-to-let developments over fifty units (DECLG, 2015). It was argued that a demand for this type of housing was created following the prohibition of 'bedsits' in 2008. Dublin City Council and Cork had already introduced slightly higher-size standards. However, it is likely that these will be reduced, in line with national standards, as the construction industry continues to claim that they are not commercially viable. A similar argument was made by the industry in the UK in response to 2007 proposals for increased space standards. (Space standards there are near the bottom of the European range. See Brady and Roche [2014: 226].) Given the aforementioned research revealing that inadequate space can be a trigger for people to relocate from apartments to suburban houses, amendments to these standards needs to be monitored carefully. In addition, other forms of higher-density, high-quality housing need to be increased in urban areas (e.g. mews, terraced, two- or three-family dwellings).

With the economic recovery, rising employment and continued population increase, there is now a shortage of affordable housing in cities and a need to increase supply there. Increasing the supply of social housing during the GFC and early years of recovery was extremely difficult for a variety of reasons (see NESC (2014b) for a detailed discussion). These include the model of social housing finance, which differs from that used in many European countries (Cahill, 2014: 23–25). In the mid-1980s, the model changed from using borrowed finance to 100 per cent capital grants from government. When government capital spending was cut (in the 1980s and again after 2008), new social housing construction was significantly reduced. In addition, below-cost rents detract from the financial sustainability of social housing. Cost-based rents, where rents cover the actual costs of providing and maintaining dwellings, would be too high for typical social housing tenants at present and policies to expand cost-renting have been recommended but the benefits of this approach will take some time to emerge (NESC, 2014b: xi; DECLG, 2014). The 2014 Social Housing Strategy aims to provide 35,000 additional housing units (built or acquired) by local authorities and AHBs, with another 75,000 households accommodated in the private rented sector (DECLG, 2014: 14). Phase one (2015–2017) focuses on exchequer-funded construction and acquisition. Considerable progress has been made (DECLG and the Housing Agency, 2016), but there is a long way to go to meet housing need. Due to their low rental incomes, local authorities are categorised in the general government sector for EU fiscal policy rules and borrowing for social housing construction adds to the overall deficit and debt. Following the introduction of the new European rules limiting capital expenditure of government, phase two (2018–2020) will concentrate on 'off-balance sheet' funding with private finance raised via special funding vehicles and public–private partnerships. Off-balance sheet borrowing and onward lending to AHBs will help to address the financial issue for them. Since 2008, the funding model was revised for them and borrowed finance reintroduced, consisting of low interest loans at preferential rates. As rents in this sector continue to be low, AHBs receive an ongoing subsidy of approximately 92 per cent of the market rent, which can result in them receiving more than 100 per cent of the market rent. This is likely to expand provision in this sector. However, there is a need for capacity building in AHBs to enable them to become large-scale developers (NESC, 2014b). It should be noted, however, that moves to increase social housing are undermined by the continued promotion of tenant purchase schemes for local authority housing.

Insufficient housing in Dublin has led to the establishment of a task force to coordinate supply there. It argues that key impediments to increased supply are structural and financial issues within the housing and development sector, including problems of access to development finance and the capacity of the construction industry. In 2015 a number of measures were introduced to stimulate supply in central urban areas, including a new vacant site levy, and reduced development contributions in locations with unsold units. These measures may help to incentivise development on underutilised sites when they come into effect but the levy may not be sufficiently large given current house price inflation. Under planning law, up to 20 per cent of residentially zoned land could be used for social and affordable housing, and the 2014 Social Housing Strategy notes this will ensure that social housing is provided alongside private housing as private construction picks up. In 2015, this requirement was reduced from up to 20 per cent to 10 per cent for social housing. An option for developers to make a payment instead of providing social housing was removed, but the social housing can still be provided at an alternative location. However, local authorities and AHBs face considerable problems from residents pressurising them to restrict housing development on vacant sites, especially for social housing (Winston, 2008).

As increasing housing supply will take time, the spotlight has turned to addressing some issues in the private rented sector, especially affordability, given the substantial rent inflation, especially in Dublin. In 2015, the government introduced greater rent certainty via biennial rather than annual rent reviews and a requirement that tenants be given 90 days' notice of a change in rent. Despite calls to link rent increases (or decreases) during a tenancy to the annual percentage change in the consumer price index (e.g. Threshold, 2015), the option chosen merely requires landlords not to charge more than the 'market rent' for the area. A new tax relief scheme was introduced to encourage landlords to rent to those in receipt of social housing supports for a three-year period.

Affordable home ownership is still a considerable problem, as revealed by the figures on mortgage arrears (see above), but some progress has been made. As one of the reasons for arrears among home owners was unemployment (Norris and Brooke, 2011), rising employment is an important driver of this progress. However, mortgage unaffordability during the Celtic Tiger era also explains the arrears situation of many households (NESC, 2014b). It is hoped that more progress will be achieved via: improved personal insolvency legislation; Central Bank work with lenders to implement sustainable solutions for distressed borrowers by agreed deadlines (Mortgage Arrears Resolution Targets – MART); mortgage-to-rent and mortgage-to-lease arrangements; and new legislation on repossessions. Resolving the situation will take considerable time and effort. Long-term arrears on buy-to-let mortgages are also an important challenge here as research on the increase in homeless families has indicated that a proportion arise from situations where receivers take possession of a property and terminate tenancies (Walsh and Harvey, 2015). To stem future problems with unsustainable mortgages, in 2015 the Central Bank introduced new loan-to-income restrictions on residential mortgage lending of 3.5 times gross annual income for all new lending to purchase principal-dwelling homes. This, of course, means that affordable and sustainable housing options need to be expanded, especially for 'intermediate households' (NESC, 2014a: xi).

Addressing energy poverty

Energy poverty is a multi-dimensional problem, including environmental (e.g. energy efficiency and emissions), social and economic issues (poverty and social inclusion). It is also an important social policy issue because increasingly it is linked to a range of health problems (Baker, 2001; Barnes *et al.*, 2008; IPHI, 2009; Liddell and Morris, 2010; O'Meara, 2015; Romero-Ortuno *et al.*, 2013; Scott *et al.*, 2008). Given its complexity, a range of strategies is required to address it. As it is highly correlated with basic deprivation (Watson and Maitre, 2015), improving the economic situation of households is essential. However, addressing fluctuating and high energy costs and improving the energy efficiency of housing are also important. A number of measures have been introduced in response to EU directives and regulations, climate change targets and national strategies. They include: improving building regulations; grants to home owners to improve their dwellings or to assist social welfare recipients with the cost of their fuel; and education campaigns to increase knowledge/skills among relevant stakeholders. Residential energy consumption improved significantly between 2006 and 2011, but SEAI (2014) argues that research is required to establish definitively the relative contributions of each of these approaches, as well as changing energy prices (including carbon tax) and household incomes. However, it maintains that improved building regulations and some retrofitting schemes have had more impact than other measures (SEAI, 2014: 80).

Building regulations

In 2007, the thermal efficiency of Irish dwellings was among the lowest in Europe (McAvoy, 2007). Thermal performance standards were not introduced in Ireland until 1979 and 44 per cent of residential dwellings were built before 1980 (SEAI, 2010). However, a number of revisions to building regulations since 1992 have improved the energy efficiency of new dwellings. Increased insulation standards were introduced in 1991 but, from an energy perspective, the most important changes have been introduced since 2002. It is estimated that these have significantly improved insulation, ventilation, heating systems and the overall energy performance requirements of new housing (SEAI, 2014). In particular, Part L of the regulations (Conservation of Fuel and Energy) aims to promote progress towards low to zero carbon buildings. Regulations introduced in 2008 set an energy performance target for new dwellings of a 40 per cent improvement on the 2002 standards, while those introduced in 2011 amended this to a 60 per cent reduction. In addition, the 2008 regulations stipulated that all new dwellings must have some contribution from renewable energy and, since 2008, replacement oil or gas boilers must be condensing versions of the same, 'where practical'. New Irish regulations for near zero energy homes were expected at the time of writing, which would result in a further of 25 per cent reduction on the 2011 regulations, and the energy required should – to a significant extent – be from renewable sources.

These changes have resulted in some positive developments. An analysis of BER certificates indicates that energy ratings have improved (SEAI, 2014: 60). Data on BERs for residential dwellings sold, rented or upgraded since 2007 reveal that the most common rating was D1 or C3, with the average being C1 (SEAI, 2014: 61). However, in 2008, approximately three-quarters of all new houses had a BER of B3 or better, and since 2010 the share of A- or B-rated new houses has increased to over 90 per cent, although very few houses were built in these latter years. One study reveals that an improved BER rating is associated with reduced household energy expenditure and, for a significant proportion of households, the savings are sufficient to fund the investment needed to produce the improvement (Curtis and Pentecost, 2015).

There are some limitations to the impact of building regulations. First, research from other countries reveals that 'rebound effects' may occur with improved thermal standards, resulting in lower energy savings than expected (Priemus, 2005). Examples include residents choosing higher comfort levels (longer heating hours, heating more rooms) and/or larger dwellings (SEAI, 2014: 50). However, this may result in positive health impacts, especially for older people, as in Ireland they tend to have more housing quality issues (Nolan and Winston, 2011). Second, improving building regulations is essential but compliance depends on competent on-site supervision, and some studies have revealed problems in this regard in the past (Winston, 2008). In 2014, tighter building controls were introduced to address inspection and enforcement issues in the construction sector (Hayden, 2014b). In 2015, these were amended for one-off houses and self-building, a move which has been described as a 'dilution of the amendments' (Engineers Ireland, 2015). Another concern is that the construction industry is lobbying to reduce the energy standards and passive house requirements on the grounds of commercial viability. These trends may impede Ireland's progress towards near zero energy homes.

Retrofitting programmes

Retrofitting programmes to improve the existing housing stock are essential, given that almost half of the stock was built before energy specific building requirements were introduced (SEAI, 2014). A number of schemes have operated via the Housing Ministry and local authorities (e.g. Housing Aid for Older People living in private housing; and the energy efficiency retrofit and central heating programmes for local authority dwellings). In addition, the SEAI has run a series of schemes with funding from national government. This section focuses on the Warmer Homes Scheme, established in 2000 to improve the energy efficiency and comfort levels of fuel poor households. Now subsumed under the Better Energy Homes Scheme, it funds attic insulation, draught proofing, lagging jackets, energy efficient lighting, cavity wall insulation and energy advice. The work is carried out by non-profit organisations and private contractors. Participating households must be in receipt of the means-tested National Fuel Allowance Scheme. Research into the scheme revealed that thermal energy consumption decreased per dwelling (SEAI, 2014: 67). There are likely to be rebound effects as residents seek more comfort, but 'the social and health benefits arising from increased household comfort and convenience are as important as the overall energy savings for low income households' (SEAI, 2014: 67).

More recently, SEAI has introduced an area-based programme within the Better Energy Homes Scheme to target low-income households at risk of energy poverty living in mixed ownership estates. Employing a partnership approach (local authorities, housing associations, community organisations, energy suppliers and contractors), it aims to stimulate employment via sustainable energy upgrading projects. Seventy per cent of homes in each project must be energy poor, and homes are considered to be at risk of energy poverty if the owner/tenant is in receipt of certain social welfare payments. The dwelling must have been built before 2002, although those built between 2002 and 2006 are also considered. Energy poor homes owned privately, by housing associations or by charities receive funding for up to 100 per cent of eligible costs while energy poor local authority dwellings receive up to 55 per cent of costs. Eligible measures include: roof and/or wall insulation; windows; external doors; boiler upgrade/replacement; heating controls upgrade; solid fuel room heaters; solar panels; chimney draught excluders; and heat pumps. An evaluation of this scheme is pending but, given the results of earlier evaluations, it is likely to prove a cost-effective, efficient approach to alleviating fuel poverty for low-income households while also improving essential knowledge and skills (see, for example, Scheer *et al.*, 2013).

These schemes have some limitations, including: they do not provide a solution for all vulnerable households; additional work will be required over time to reduce inefficiencies over the longer term; and there is an inadequate focus on renewable energy in the grant-aided schemes. Another important weakness is that the schemes ignore private renters. The latest energy poverty strategy (DCENR, 2015a) indicates that the government intends to address this issue.

Conclusions

This chapter has focused on a number of challenges impeding the transition to more sustainable urban communities in Ireland. These include issues relating to land use planning, settlement and commuting patterns, particularly the inappropriate location of housing; high levels of commuting by private car, residential energy consumption and energy poverty; and housing affordability in cities. Progress has been made on some issues. For the most part, there is good policy and guidance regarding sustainable housing but it is essential that this is implemented. There is some scope for further urban densification, not just in the inner city but also in the inner and outer suburbs. Increasing the diversity of the built form could accommodate a greater variety of household types, including families, in urban areas as long as there is adequate space and the dwellings and neighbourhoods are of high quality. Otherwise, city and apartment dwellers will continue to view these locations as stepping stones to detached/semi-detached dwellings in the suburbs and beyond. Increasing densification, while maintaining high-quality green space for recreation, will facilitate improved, financially viable public transport. While housing quality is relatively good overall, more progress is required to address relatively high levels of: fossil fuel dependence; energy consumption; and energy poverty among certain groups. Improved building regulations and retrofitting schemes will help to address these issues. With regard to retrofitting, there needs to be more focus on the rented sector, especially private renting, and on shifting from fossil fuels to renewable energy. With regard to the latter, more emphasis could be placed on district/estate renewable energy schemes and on grants for renewable energy. These would help increase the security of supply, reduce emissions and decrease energy poverty.

Increasing the supply of affordable urban housing is essential, especially in the cities, where particular groups face very significant affordability issues. More compact settlements, including more variety in the built form, will assist with this. Policies to expand cost-renting are most important but the benefits of this will take time to emerge. The very significant public debt has impeded progress on social housing despite recent progress, but increasing the capacity of AHBs should greatly assist with increasing the supply of social housing, given the availability of finance for that sector. 'Part V' type arrangements – where social and affordable housing are included within new housing developments – will also assist in meeting urban housing need.

Some of the issues described in this chapter reflect tensions between the economic, social and environmental dimensions of sustainable development. For example, construction industry attempts to reduce space standards and energy regulations may help reduce housing provision costs in the short term but raise serious questions about the extent to which this will create more problems for the future (e.g. energy poverty and emissions) and/or repeat the mistakes of the past by building housing that is not considered suitable or desirable over a large part of the life-cycle. To this end, more consideration could be given to life-cycle/ adaptable housing. But it also highlights that more attention should be given to good urban planning, and making the city and its neighbourhoods more attractive for long-term residence by all types of households. NESC's (2014b: ix) call for direct public policy influence on housing supply and urban development and the institutional mechanism to achieve this should be commended in this regard. Efforts across all of these domains would help with the development of more sustainable urban communities in Ireland.

References

ADEME (2015) *Energy Efficiency Trends and Policies in the Household and Tertiary Sectors*. Available at www.odyssee-mure.eu/publications/br/energy-efficiency-in-buildings.html, accessed 6 July 2016.

Arts, W. and Gelissen, J. (2010) 'Models of the welfare state', in F. Castles, S. Leibfried, J. Lewis, H. Obinger and C. Pierson (eds) *The Oxford Handbook of the Welfare State*, pp. 569–584. Oxford: Oxford University Press.

Baker, W. (2001) *Fuel Poverty and Ill Health: A Review*. Bristol: Centre for Sustainable Energy. Available at www.cse.org.uk/pdf/pub11.pdf, accessed 6 July 2016.

Barnes, M., Butt, S. and Tomaszewski, W. (2008) *The Dynamics of Bad Housing: The Impact of Bad Housing on the Living Standards of Children*. London: National Centre for Social Research.

Bouzarovski, S. (2011) *Energy Poverty in the EU: A Review of the Evidence*. Available at http://ec.europa.eu/regional_policy/archive/conferences/energy2011nov/doc/papers/bouzarovski_eu_energy_poverty_background%20paper.pdf, accessed 6 July 2016.

Brady, N. and Roche, J. (2014) 'Fit for renting: an investigation into space standards of Irish housing', in L. Sirr (ed.) *Renting in Ireland: The Social, Voluntary and Private Sectors*, pp. 222–240. Dublin: IPA.

Buckley, C. (2012) 'Implementation of the EU Nitrates Directive in the Republic of Ireland: a view from the farm', *Ecological Economics*, 78: 29–36.

Cahill, N. (2014) *Financing of Social Housing in Selected European Countries*. NESC Secretariat Paper No. 11. Dublin: National Economic and Social Council.

Callan, T., Nolan, B., Keane, C., Savage, M. and Walsh, J. (2013) *Crisis, Response and Distribution Impact: The Case of Ireland*. Economic and Social Research Institute Working Paper No. 456. Dublin: Economic and Social Research Institute.

Central Bank of Ireland (2015) *Residential Mortgage Arrears and Repossessions Statistics*. Available at www.centralbank.ie/polstats/stats/mortgagearrears/Pages/releases.aspx, accessed 6 July 2016.

Central Statistics Office (CSO) (2007) *Census 2006: Principal Demographic Results*. Dublin: Stationery Office.

Central Statistics Office (CSO) (2011) *Profile 4: The Roof over our Heads.* Available at www. cso.ie/en/media/csoie/census/documents/census2011profile4/Profile4_The_Roof_over_ our_Heads_entire_doc.pdf, accessed 6 July 2016.

Central Statistics Office (CSO) (2012a) *Population and Migration Estimates, April 2012.* Dublin: Central Statistics Office.

Central Statistics Office (CSO) (2012b) *This is Ireland: Highlights from Census 2011, Part 1.* Dublin: Central Statistics Office.

Central Statistics Office (CSO) (2015) *Survey on Income and Living Conditions: 2014 Results.* Available at www.cso.ie/en/releasesandpublications/er/silc/surveyonincome andlivingconditions2014/, accessed 6 July 2016.

Central Statistics Office (CSO) (2016) *Monthly Unemployment Statistics, February 2016.* Available at www.cso.ie/en/releasesandpublications/er/mue/monthlyunemployment february2016/, accessed 6 July 2016.

Cox, W. and Pavletch, H. (2013) *10the Annual Demographia International Housing Affordability Survey 2014: Ratings for Metropolitan Markets.* Available at www. demographia.com/dhi.pdf, accessed 6 July 2016.

Curtis, J. and Pentecost, A. (2015) 'Household fuel expenditure and residential building energy efficiency ratings in Ireland', *Energy Policy*, 76: 57–65.

Daly, M. and Yeates, N. (2003) 'Common origins, different paths: adaptation and change in social security in Britain and Ireland', *Policy and Politics*, 31(1): 85–97.

Davies, A. (2013) 'When clean and green meet the Emerald Isle', in M. Zapata and M. Hall (eds) *Organising Waste in the City: International Perspectives on Narratives and Practices*, pp. 63–82. Bristol: Policy Press.

Department of Communications, Energy and Natural Resources (DCENR) (2007) *Delivering a Sustainable Energy Future for Ireland.* Dublin: Department of Communications Energy and Natural Resources.

Department of Communications, Energy and Natural Resources (DCENR) (2009) *National Energy Poverty Strategy.* Dublin: Department of Communications Energy and Natural Resources.

Department of Communications, Energy and Natural Resources (DCENR) (2011) *Warmer Homes: A Strategy for Affordable Energy in Ireland.* Dublin: Department of Communications Energy and Natural Resources. Available at www.dcenr.gov.ie/NR/rdonlyres/53F3AC25- 22F8-4E94-AB73-352F417971D7/0/AffordableEnergyStrategyFINAL.pdf, accessed 6 July 2016.

Department of Communications, Energy and Natural Resources (DCENR) (2015a) *A Strategy to Combat Energy Poverty 2016–2019.* Dublin: Department of Communications, Energy and Natural Resources.

Department of Communications, Energy and Natural Resources (DCENR) (2015b) *Ireland's Transition to a Low Carbon Energy Future 2015–2030.* Dublin: Department of Communications Energy and Natural Resources.

Department of the Environment, Community and Local Government (DECLG) (2011a) *Housing Policy Statement.* Dublin: Department of the Environment, Community and Local Government.

Department of the Environment, Community and Local Government (DECLG) (2012a) *Towards Nearly Zero Energy Buildings in Ireland Planning for 2020 and beyond.* Dublin: Department of Environment, Community and Local Government.

Department of the Environment, Community and Local Government (DECLG) (2014) *Social Housing Strategy 2020.* Dublin: Department of Environment, Community and Local Government.

Department of the Environment, Community and Local Government (DECLG) (2015) *Information Note on 2015 Apartment Guidelines.* Dublin: Department of Environment, Community and Local Government.

Department of the Environment, Community and Local Government (DECLG) and the Housing Agency (2016) *Social Housing Output in 2015.* Dublin: Department of Environment, Community and Local Government and the Housing Agency.

Department of the Environment, Heritage and Local Government (DEHLG) (2007a) *Building Regulations 2007 – Technical Guidance Document L – Conservation of Fuel and Energy – Dwellings.* Dublin: Department of the Environment, Heritage and Local Government.

Department of the Environment, Heritage and Local Government (DEHLG) (2007b) *Delivering Homes Sustaining Communities: Statement on Housing Policy.* Dublin: Department of the Environment, Heritage and Local Government.

Department of the Environment, Heritage and Local Government (DEHLG) (2007c) *Quality Housing for Sustainable Communities: Best Practice Guidelines for Delivering Homes, Sustaining Communities.* Dublin: Department of the Environment, Heritage and Local Government.

Department of the Environment, Heritage and Local Government (DEHLG) (2008a) *Sustainable Residential Development in Urban Areas.* Dublin: Department of the Environment, Heritage and Local Government.

Department of the Environment, Heritage and Local Government (DEHLG) (2008b) *Urban Design Manual: A Best Practice Guide.* Dublin: Department of the Environment, Heritage and Local Government.

Department of the Environment, Heritage and Local Government (DEHLG) (2009a) *Sustainable Residential Development in Urban Areas.* Dublin: Department of the Environment, Heritage and Local Government.

Department of the Environment, Heritage and Local Government (DEHLG) (2009b) *Urban Design Manual: A Best Practice Guide.* Dublin: Department of the Environment, Heritage and Local Government

Department of the Environment, Heritage and Local Government (DEHLG) and Office of Public Works (OPW) (2009) *The Planning System and Flood Risk Management: Guidelines for Planning Authorities.* Dublin: Department of the Environment, Heritage and Local Government.

Demographia (2016) *12th Annual Demographia International Housing Affordability Survey 2016: Rating Middle Income Housing Affordability.* Available at www. demographia.com/dhi.pdf, accessed 6 July 2016.

Dol, K. and Haffner, M. (2010) *Housing Statistics in the European Union 2010.* The Hague: Ministry of Interior and Kingdom Relations.

Doyle, N. (2009) 'Housing finance developments in Ireland', *Central Bank Quarterly Bulletin,* 9: 75–88.

Dukelow, F. (2011) 'Economic crisis and welfare retrenchment: comparing Irish policy responses in the 1970s and 1980s with the present', *Social Policy and Administration,* 45(4): 408–429.

Engineers Ireland (2015) *Engineers Ireland Respond to the Reforms of the Building Control (Amendment) Regulations S.I. No. 9 of 2014.* Available at www.engineersireland.ie/ Communications/Press-Archive/Engineers-Ireland-response-to-the-reforms-of-the-B. aspx, accessed 6 July 2016.

European Commission (2015) *Country Report Ireland 2015,* COM (2015) 85 final. Brussels: European Commission. Available at http://ec.europa.eu/europe2020/pdf/ csr2015/cr2015_ireland_en.pdf, accessed 6 July 2016.

Fahey, T. (2002) 'The family economy in the development of a welfare regime: a case study', *European Sociological Review*, 18(1): 51–64.

Fahey, T. (ed.) (1999) *Social Housing in Ireland*. Dublin: IPA.

FitzGerald, E. and Winston, N. (2011) 'Housing, equality and inequality', in M. Norris and D. Redmond (eds) *Housing Contemporary Ireland*, pp. 224–244. Dordrecht: Springer.

Flynn, B. (2007) *The Blame Game*. Dublin: Irish Academic Press.

Grist, B. (2012) *An Introduction to Irish Planning Law*. Second edition. Dublin: IPA.

Hayden, A. (2014a) 'Local authority rented housing: the path to decline', in L. Sirr (ed.) *Renting in Ireland*, pp. 107–120. Dublin: IPA.

Hayden, R. (2014b) 'New building regulations', *Housing Ireland*, 7: 30–33.

Hick, R. (2014) 'From Celtic Tiger to crisis: progress, problems and prospects for social security in Ireland', *Social Policy and Administration*, 48(4): 394–412.

Housing Agency and DECLG (2015) *Resolving Unfinished Housing Developments: Annual Progress Report on Actions to Address Unfinished Housing Developments*. Dublin: Housing Agency and DECLG.

Howley, P., Scott, M. and Redmond, D. (2009) 'An examination of residential preferences for less sustainable housing: exploring future mobility among Dublin central city residents', *Cities*, 26(1): 1–8.

Institute for Public Health in Ireland (2009) *Annual Update on Fuel Poverty and Health 2009*. Available at www.publichealth.ie/files/file/Annual%20Update%20on%20Fuel%20Poverty%20and %20Health%202009.pdf, accessed 6 July 2016.

International Monetary Fund (IMF) (2015) *Ireland: Fourth Post-program Monitoring Discussions*. Washington, DC: IMF.

Kelly, M. (2009) *The Irish Credit Bubble*. UCD Centre for Economic Research Working Paper 09/32. Dublin: University College Dublin.

Kemeny, J. (1995) *From Public Housing to the Social Market: Rental Policy Strategy in Comparative Perspective*. London: Routledge.

Kennedy, K., Giblin, T. and McHugh, D. (1988) *The Economic Development of Ireland in the Twentieth Century*. London: Routledge.

Kitchin, R., O'Callaghan, C. and Gleeson, J. (2014) 'The new ruins of Ireland? Unfinished estates in the post-Celtic Tiger era', *International Journal of Urban and Regional Research*, 38(3): 1069–1080.

Liddell, C. and Morris, C. (2010) 'Fuel poverty and human health: a review of recent evidence', *Energy Policy*, 38: 2987–2997.

McAvoy, H. (2007) *All Ireland Poverty Paper on Fuel Poverty*. Dublin: Institute of Public Health in Ireland. Available at http://lenus.ie/hse/bitstream/10147/45784/1/9138.pdf, accessed 6 July 2016.

McCashin, T. (2012) 'Social security expenditures in Ireland, 1981–2007', *Policy and Politics*, 40(4): 547–567.

Mahon, A., Flood, F., Faherty, M. and Keys, G. (2012) *The Final Report of the Tribunal of Inquiry into Certain Planning Matters and Payments*. Dublin: Stationery Office.

Mahon, M. and O'Cinneide, M. (2010) 'Housing supply and residential segregation in Ireland', *Urban Studies*, 47: 2983–3012.

Motherway, D., Kelly, M., Faughnan, P. and Tovey, H. (2003) *Trends in Irish Environmental Attitudes between 1993 and 2002*. Dublin: Social Science Research Centre and Department of Sociology, University College Dublin, and Department of Sociology, Trinity College Dublin.

National Economic and Social Council (NESC) (2014a) *Homeownership and Rental: What Road is Ireland on?* Dublin: National Economic and Social Council.

National Economic and Social Council (NESC) (2014b) *Social Housing at the Crossroads: Possibilities for Investment, Provision and Cost Rental*. Dublin: National Economic and Social Council.

Nolan, B. and Winston, N. (2011) 'Dimensions of housing deprivation for older people in Ireland', *Social Indicators Research*, 104: 369–385.

Nolan, B., Whelan, C., Calvert, E., Healy, D., Fahey, T., Mulcahy, A., Norris, M. and Winston, N. (2014) 'Ireland and its impact in boom and bust', in W. Salverda, B. Nolan, D. Checci, I. Marx, A. McKnight, I. Toth *et al.* (eds) *Changing Inequalities and Societal Impacts in Thirty Rich Countries*, pp. 346–368. Oxford: Oxford University Press.

Norris, M. (2014a) 'Path dependence and critical junctures in Irish rental policy: from dualist to unitary rental markets?', *Housing Studies*, 29(5): 616–637.

Norris, M. (2014b) 'Policy drivers of private renting', in L. Sirr (ed.) *Renting in Ireland*, pp. 19–37. Dublin: IPA.

Norris, M. and Brooke, S. (2011) *Lifting the Load: Help for People in Mortgage Arrears*. Dublin: Citizens Information Board.

Norris, M. and Coates, D. (2014) 'How housing killed the Celtic Tiger: anatomy and consequences of Ireland's boom and bust', *Journal of Housing and the Built Environment*, 29(2): 299–315.

Norris, M. and Domanski, H. (2009) 'Housing conditions, states, markets and households', *Journal of Comparative Public Policy Analysis*, 11(3): 385–407.

Norris, M. and Winston, N. (2003) *Housing Policy Review 1990–2002*. Dublin: Stationery Office.

Norris, M. and Winston, N. (2009) 'Rising second home numbers in rural Ireland: distribution, drivers and implications', *European Planning Studies*, 17(9): 1303–1322.

Norris, M. and Winston, N. (2011) 'Transforming Irish home ownership through credit deregulation, boom and crunch', *International Journal of Housing Policy*, 11(1): 1–21.

O'Brien, L. and Dillon, B. (1982) *Private Rented: The Forgotten Sector*. Dublin: Threshold.

ODYSSEE (n.d.) Database. Available at www.indicators.odyssee-mure.eu/online-indicators.html, accessed 6 July 2016.

Office of the Deputy Prime Minister (ODPM) (2005) *Sustainable Communities: People, Places, and Prosperity* (Bristol Accord). London: ODPM.

O'Meara, G. (2015) 'A review of the literature on fuel poverty with a focus on Ireland', *Social Indicators Research*, DOI 10.1007/s11205-015-1031-5.

Priemus, H. (2005) 'How to make housing sustainable? The Dutch experience', *Environment and Planning B*, 32: 15–19.

Radl Rogerova, D., Hana, P. and Novak, P. (2014) *Property Index Overview of European Residential Markets*. Deloitte. Available at www2.deloitte.com/be/en/pages/real-estate/articles/be-european-residential-property-index-2015.html, accessed 6 July 2016.

Romero-Ortuno, R., Tempany, M., Dennis, L., O'Riordan, D. and Silke, B. (2013) 'Deprivation in cold weather increases the risk of hospital admission with hypothermia in older people', *Irish Journal of Medical Science*, 182: 513–518.

Scheer, J., Clancy, M. and Ní Hógáin, S. (2013) 'Quantification of energy savings from Ireland's Home Energy Savings scheme: an ex-post billing analysis', *Energy Efficiency*, 6: 25–48.

Scott, S., Lyons, S., Keane, C., McCarthy, D. and Tol, R. (2008) *Fuel Poverty in Ireland: Extent, Affected Groups and Policy Issues*. Working Paper No. 262. Dublin: Economic and Social Research Institute. Availabe at www.esri.ie/UserFiles/publications/20081110114951/WP262.pdf, accessed 6 July 2016.

Sustainable Energy Authority of Ireland (SEAI) (2010) *Residential Energy Roadmap 2010–2050*. Dublin: Sustainable Energy Authority of Ireland.

Sustainable Energy Authority of Ireland (SEAI) (2014) *Energy in Ireland 1990–2013*. Dublin: Sustainable Energy Authority of Ireland.

Threshold (2015) *Rent Certainty*. Dublin: Threshold.

Tribunal of Inquiry into Certain Planning Matters and Payment (2012) *The Final Report of the Tribunal of Inquiry into Certain Planning Matters and Payments* (Mahon Report). Dublin: Stationery Office.

Walsh, K. and Harvey, B. (2015) *Family Experiences of Pathways into Homelessness: The Families' Perspective*. Dublin: Housing Agency.

Watson, D. and Maitre, B. (2015) 'Is fuel poverty in Ireland a distinct type of deprivation?', *Economic and Social Review*, 46(2): 267–291.

Wickham, J. (2006) 'Public transport systems: the sinews of European urban citizenship', *European Societies*, 8(1): 3–26.

Whelan, C. and Maitre, B. (2014) 'The Great Recession and the changing distribution of economic vulnerability by social class: the Irish case', *Journal of European Social Policy*, 24(5): 470–485.

Williams, B., Hughes, B. and Shiels, P. (2007) *Urban Sprawl and Market Fragmentation*. Dublin: Society of Chartered Surveyors.

Williams, B. and Shiels, P. (2000) 'Acceleration into sprawl: causes and potential policy responses', *Quarterly Economic Commentary*, June: 37–73.

Winston, N. (2008) *Urban Regeneration for a Sustainable City: The Role of Housing*. Available at http://researchrepository.ucd.ie/bitstream/handle/10197/3906/Blue_book.pdf?sequence=1, accessed 6 July 2016.

3   Spain

*Montserrat Pareja-Eastaway and
María-Teresa Sánchez-Martínez*

Introduction

Spain approved its current constitution in 1978. This led to the decentralization of policies, income sources and responsibilities to lower levels of government: 17 Autonomous Communities or regional government (with varying degrees of power), 52 provinces and 8,114 municipalities which constitute the three main tiers of government in Spain. Today, the multilevel governance that characterizes the Spanish government is under consideration given the serious financial imbalances displayed by both the central government and the Autonomous Communities and municipalities.

Approximately 46.5 million people live in Spain, making it the fifth most populous country in Europe after Germany, France, Italy and the UK. Half of the population lives in densely populated areas and about one-third in rural areas, which geographically account for 90 per cent of the territory (Camarero, 2009). Demographic, labour and social changes are reflected in the forms of cohabitation, in such a way that household size is increasingly smaller and household composition increasingly diverse (PWC, 2015).

Spain underwent a delayed urbanization process compared with other European countries that have also experienced recent metropolitan development. In general terms, a duality is detected in the country: on the one hand, the significant dynamism of large urban areas such as Madrid and Barcelona and, on the other, progressive rural depopulation and rural crisis (Ministry of Environment, 2011). Data on average population density in Spain (91.4 inhabitants/km^2 in 2008) shows that 70 per cent of the population lives in large urban areas. The city model that exists throughout Spain is the Mediterranean compact city model, adapted to the diversity and complexity of the territory: the preservation of the foundational elements of this model is considered a crucial factor in the design of urban and local sustainability for the future (Ministry of Environment, 2011). In fact, the mixed nature of uses and functions has to a large degree characterized Spanish cities, guaranteeing sustainable development and good social cohesion (LA21, 2008). Thus, the relatively good performance of Spain on the EPI (79.79; see Table 1.1) might be related to the huge uninhabited territory, with more guarantees for preservation, given the components of the index, as well as due to the reasons attributed to the characterization of the Mediterranean city.

GDP per capita in 2013 was €22,500, which is below the EU28 average. Of the countries analysed in this book, only Hungary and Romania had a lower GDP in that year (see Table 1.1). For years, Spain was among the poorest EU countries, together with Greece, Ireland and Portugal, and benefited from a broad range of EU policies aimed at achieving greater territorial cohesion. When Central and Eastern European countries joined the EU in 2001, Spain became a relatively prosperous member state, a net contributor to EU funds, and received fewer subsidies (Piedrafita *et al.*, 2006).

Spain's experience of the Global Financial Crisis (GFC) has been particularly severe. Over the period 2005–2009, growth rates were above the EU15 and the EU28 averages. However, since 2010, Spain has fallen far behind in terms of GDP growth rates. Since 2014, growth rates have improved (1.8 per cent in 2014 and 3.2 per cent in 2015) due to the international recovery, the positive performance of the tourist sector, rising exports and declining energy prices. However, the labour market is currently one of the biggest problems. The unemployment rate was 23.7 per cent in 2014 and 20.9 per cent in 2015, and for young people (under 25 years of age) it was over 51 per cent and 46.2 per cent, respectively (INE (*EPA*), 2014, 2015).

During the expansionary period (1996–2007), Spain fostered 'bricks and mortar'-based growth that was environmentally irresponsible and extremely vulnerable to exogenous shocks, such as the GFC. Residential construction was rampant even in natural areas due to the high profitability the real estate sector offered during the period. This exploitation of land has led Spain towards serious problems with environmental sustainability, primarily affecting large cities and tourist areas, especially those with many second homes (Delgado Viñas, 2008; Valenzuela, 2008). In addition, this phenomenon increased urban sprawl, with a greater tendency towards unsustainability given, for example, the rise in commuting, the costs of energy distribution in an expanded city, and the socio-spatial segregation of residents. Reflection and analysis in academic circles that reconsiders the desirable target rate of city building, comparing the idea of the compact city versus the extended city, has been developed, but there has been relatively little practice on implementing laws that would guarantee sustainability (Aguado and Etxebarria, 2003; Indovina, 2007; Molini and Salgado, 2010).

Spain is regarded as an example of a 'clientelist', 'familiarist' or Mediterranean welfare system (Allen *et al.*, 2004; Gal, 2010). Since the GFC, there has been a dramatic reduction in public spending, including severe cuts in social protection, which has pushed the welfare state back to what it was in the 1980s (Navarro *et al.*, 2012). The socio-economic effects of the crisis in Spain have been devastating, including increased risk of poverty, a shrinking of the middle classes and wage cuts, among other issues. Social exclusion has increased, and in 2013 housing and employment were the dimensions in which most deprivation was experienced (Alguacil *et al.*, 2015). Rising levels of social exclusion and deprivation have led to a number of grassroots movements for change, including examples of citizens seeking access to decent housing and attempting to counterbalance the fragility and vulnerability they face.

The housing system in Spain

The Spanish housing system is an example of tenure imbalance where homeownership dominates and there is a lack of social housing (Pareja-Eastaway and Sánchez-Martínez, 2014). While in 2004 homeownership represented 80 per cent and the private rented sector 10 per cent, by 2014 private renting had increased to 15 per cent while homeownership had dropped to 78 per cent. The dominance of homeownership results in a set of significant socio-demographic realities in Spain: the delayed emancipation of young people and the consequent postponement of the formation of their first home due to a shortage of reasonably priced rental housing (Jurado Guerrero, 2004; Pareja-Eastaway, 2007). This delay and the high percentage of young people living as homeowners differentiates young households in Southern Europe from the European average (Módenes *et al.*, 2013; Allen *et al.*, 2004). The scarcity of private rental housing contributes to a lack of flexibility and mobility in the labour market. However, after the crisis, demand for rental housing has emerged as an alternative to the insecurity of homeownership for vulnerable families in periods of economic recession (Pareja-Eastaway and Sánchez-Martínez, 2015). This has resulted in an increase in the relative scarcity of rental accommodation and in rent inflation. The small percentage of (publicly rented) social housing (approximately 2 per cent of the total) is insufficient to meet the needs of vulnerable groups. The main reasons for the limited size of rented tenure in the Spanish housing system include problems in the management of public rented housing, difficulties defining contracts and low yields in the private sector (Burón Cuadrado, 2008).

In 2012, 65 per cent of the population lived in flats (Table 1.1), the highest percentage of all the countries considered in this volume. Over the last decade, Spain has experienced a boom in construction and unprecedented price increases: in the period 1997–2007 the increase was over 10 per cent per year, and in some years up to approximately 30 per cent (Sánchez-Martínez, 2008). After the housing bubble burst in 2008, the Spanish housing market underwent significant adjustment. From 2007 to 2014 house prices declined by approximately 40 per cent on average, ranging from 20 per cent falls in the least affected areas to over 70 per cent falls in the most deprived areas.

The model of Spanish growth since 1996 has two paradoxes that are difficult to marry. On one hand, while more housing was built in this period than in any other, it was also the period with the greatest number of homeless families (Romero, 2010; Hoekstra *et al.*, 2010). On the other hand, despite the high number of new developments, vacant housing accounts for a disproportionate part of the overall stock. In 2013, housing stock in Spain was estimated at 25,441,306 dwellings (Ministry of Public Works, 2013), with 75 per cent of housing stock used as main residences and 25 per cent as second homes. During the real estate boom period (2001–2013), stock grew by more than four million dwellings, an increase of 21 per cent over the period. However, housing *construction* was dissociated from housing *need*, including the arrival of immigrants, and very little social housing was constructed. As a result, there is a clear shortage of affordable housing close to centres of employment. At the same time, there are almost 3.5 million vacant dwellings (14 per cent of total stock), a situation which may give rise to many negative consequences (Vinuesa Angulo, 2008). From the economic point of view, these are very valuable assets that are not being utilized and which are an inefficient use of urban space. In environmental terms, land is a basic and limited good and residential development beyond housing need can reflect excessive land consumption, is detrimental to the landscape and deflects from other non-residential uses, such as recreation, agriculture and energy production.

Between 2008 and the third quarter of 2014 there were 360,042 evictions across Spain (General Council of the Judiciary, 2014). This problem is leading to a new type of 'unsafe' housing, according to the definition of the European Typology of Homelessness and Housing Exclusion (ETHOS) (Amore *et al.*, 2011). Between 2007 and 2014 approximately 600,000 foreclosures took place in Spain. Families evicted for not paying their mortgage or rent are forced to live with relatives or friends, sublet rooms in apartments, or live as squatters in empty buildings due to the lack of social rental options. Homes that are owned by the banks have often been withheld in order to avert, as far as possible, an even greater house price crash. However, there is also a stock of built social housing that remains vacant due to the lack of management tools that facilitate letting to target people (Ombudsman, 2013).

Current guidelines for housing policy (2013–2016 State Plan for the Promotion of Rented Housing, Building Rehabilitation, and Urban Regeneration and Renewal) considers private rented housing as the main target for promotion through public intervention alongside rehabilitation policies for the built environment. Issues such as the legal insecurity of landlords and the consequent difficulties in carrying out evictions are priorities in this new scenario.

Sustainable communities and housing in Spain

Definitions

The concept of 'sustainable communities', as such, is rarely used in Spain. However, a holistic view of sustainability and sustainable development of the territory can be found in the approach of the Spanish Ministry for Public Works, which is in charge of sketching the main guidelines relating to sustainability and housing. The Ministry's approach corresponds with the view of three integrated pillars of sustainability, guiding its policies towards the goal of 'contributing to the construction of cities that are more efficient in economic terms, more equitable in social terms, and more sustainable in environmental terms' (Ministry of Public Works, 2010). This approach responds to several principles established at different levels of governance, such as the international Kyoto Protocol, Article 47 of the Spanish Constitution, which establishes that 'every Spanish citizen has the right to enjoy decent and adequate housing', and various European-level conventions and agreements (e.g. the Lisbon Agenda, the Gothenburg Agenda and the European Territorial Strategy).

The Ministry has the following view of land: 'in addition to being an economic resource, [it] is one of the most valuable natural resources we have, the regulation of which requires combining a series of different factors: the environment, quality of life, energy efficiency, service provision, social cohesion' (Ministry of Public Works, 2010). This holistic approach presents an understanding of sustainable development that is linked to the idea of balanced, socially cohesive communities. In addition to principles established at the European level, the 2007 Land Law includes the 'principle of sustainable territorial and urban development', where we can also detect the goal of balancing the components of sustainability.

In 2011, the Spanish Strategy for Urban and Local Sustainability (EESUL) was produced driven by the national state administration. Employing the principles of the 2006 European Thematic Strategy on the Urban Environment and the 2007 Spanish Sustainable Development Strategy, the EESUL added principles concerning rural–city relations and climate change. It was designed with the involvement of different stakeholders beyond the relevant ministries (Housing and the Environment). These included the Urban Ecology Agency of Barcelona (BCNecologia), the Network of Local Sustainable Development Networks and its constituent regional and local networks. The EESUL aims to become a strategic non-binding framework that develops principles, goals, guidelines and measures whose effective application will allow for progress in the direction of greater urban and local sustainability (BCNecologia, 2012). However, this proposal has not yet passed before the Council of Ministers or Parliament for approval. The main obstacle to achieving commitment to it is the economic crisis, its dramatic impact on Spanish society, and the focus on short-term emergency solutions in a generalized context of public spending reductions at all levels of government.

Sustainable housing and communities' indicators

From 2005 until its closure in 2013, the Observatory of Sustainability in Spain (OSE) produced annual sustainability reports based on indicators. The use of indicators for conducting this type of analysis provided a rigorous and objective view of the situation of sustainability through comparable and reliable information. The first report, in 2005, considered 65 indicators covering economic, social and environmental dimensions. This figure increased to 155 in later reports. These indicators were intended to monitor national development strategies towards the general lines addressed by the European Community. The OSE was closed due to cutbacks in funding coming mainly from the Ministry of Agriculture and Environment and the Biodiversity Foundation. The (new) Observatory of Sustainability (OS) was set up in 2014 with the goal of bringing together relevant information on the state of sustainability at the national level, applying a set of 'verified, operational, and representative' indicators (OS, 2014: 5). Four years earlier, the Ministry of Public Works had established a holistic system of indicators for measuring progress towards sustainability in Spain's cities and towns. They covered aspects such as land use, social cohesion and the increase of biodiversity. The indicators were applied in 2011 in only four Spanish municipalities.

OS published its latest report in 2014. Sustainable indicators were measured under the assumption that GDP is not the best tool for providing information about the degree of sustainability achieved by a country, a region or a city. Nevertheless, the 'Report on the Situation of the Main Actions and Initiatives in Matters of Urban Sustainability in Spain', compiled by the group GEA 21 in 2011 for the Ministry of Public Works (Verdaguer Viana-Cárdenas, 2011), suggests that there is a fairly strong correlation between GDP and the density of sustainability initiatives. This does not hold true in only two Spanish regions: La Rioja, which has very few initiatives, despite having a high GDP; and Andalusia, where the situation is precisely the opposite. With regard to the top-ranked regions – Basque Country, Navarre, Madrid, Aragon and Catalonia – the correlation is very strong indeed, and it is only slightly weaker in the lowest-ranked regions – Extremadura, Castilla-La Mancha and Murcia.

The major concern for the OS in 2014 related to the social consequences of the economic crisis and the lack of alternatives to an economic model of development based on an expansive real estate sector. According to OS (2014), Spanish households' disposable income decreased by 14.5 per cent between 2008 and 2013. In general terms, a significant decrease in the living conditions of the Spanish population was noticeable according to OS indicators. As mentioned above, the productive model adopted during the expansion period exerted excessive pressure on the natural ecosystem and created a volatile housing market. The Spanish economy needs to reinvent itself in order to establish its pillars in a more sustainable paradigm. For instance, calculations show that the green economy in Spain might contribute to the creation of 2 million jobs until 2020, while the 1.37 million jobs that might be created through the rehabilitation of the existing housing stock deserve particular attention.

Spain has improved its performance in the use of energy but the level is still too high. With reference to primary energy, on average, 71.3 percent of energy consumption in Spain relies on imports, despite a reduction in consumption over the last decade. In a similar vein, indicators on climate change suggest the non-sustainability of Spain's development model. If we consider Spain's sustainability from an environmental point of view, a key element to bear in mind is the geoclimatic zone of the country, as well as the level of wealth and the consumption reduction strategies followed by the population. Thus, analysing electricity consumption (in kWh/dw) in 2013, according to the data drawn up by Odyssee-Mure, the highest consumption was registered by Sweden (5,704), Finland (3,255) and Austria (3,204), whereas Spain was at 2,212, close to Greece (2,292), Ireland (2,305) and Germany (2,337). With regard to consumption per dwelling for space heating (normal climate), those with the highest rates registered around 1.5 (toe/dw), such as Finland (1.58) and Denmark (1.31), while Spain, Portugal and Cyprus had the least consumption at 0.38, 0.14 and 0.27, respectively.

An interesting assessment/appraisal that Odyssee carries out is the variation of household consumption for heating between two dates. This indicator is influenced by: climatic difference between the two dates ('climatic effect'); change in the number of occupied dwellings ('more dwellings'); change in floor area of dwelling

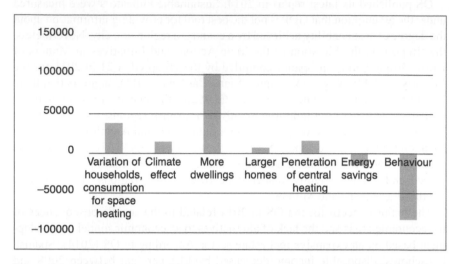

Figure 3.1 Variation of households' consumption for space heating (2000–2013) in Tj, Spain.

Source: Odyssee-Mure.

for space heating ('larger homes'); diffusion of central heating, mainly in Southern European countries; energy savings, as measured using ODEX; and other effects (mainly change in heating behaviour). In the case of Spain, if we compare 2000 and 2013, while the consumption of space heating for 2000 was 257695.7 Tj, by 2013 this had increased to 295387.1 Tj. Despite reductions in consumption due to changes in household behaviour, these do not compensate for increases in consumption provoked by increases in dwellings over the period (see Figure 3.1).

Financial vulnerability of households has increased since the GFC (Pareja-Eastaway and Sánchez-Martínez, 2016). Not only housing affordability but also households' ability to guarantee minimum standards concerning water, electricity and gas have declined over recent years. Energy poverty is understood as a household's inability to access minimum energy services for its basic needs and to maintain the home at an adequate temperature (Gordon *et al.*, 2000; Thomson and Snell, 2013), or it may be defined as any household that spends more than 10 per cent of its income on energy (Boardman, 1991). Some suggest that Portugal, Greece, Spain and Italy demonstrate the highest levels of fuel poverty in Europe (Healy and Clinch, 2002: 33; Thomson and Snell, 2013: 570). In Spain, there has been little concern about or recognition of this issue. According to the European Commission (2009: 16), 'there is no real remedial infrastructure because there was no perception of fuel poverty as a compelling social problem'. There is now rather more appreciation of the problem, yet Spain still lacks an official definition of both 'energy poverty' and 'vulnerable consumers'; indeed, in Europe, only the UK, Ireland and France have developed definitions of fuel poverty (Thomson and Snell, 2013).

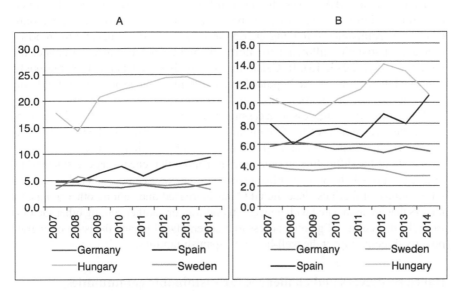

Figure 3.2 Arrears on utility bills (A) and inability to keep home adequately warm (B).

Source: Eurostat.

Figure 3.2 presents a comparative analysis of energy poverty employing two indicators: arrears on utility bills in recent months; and inability to keep the home adequately warm. We have chosen three countries to compare data with Spain: Germany, Sweden and Hungary. They cover the wide European spectrum of economic development standards and different cultures with respect to sustainability. The data reveal that the problem of energy poverty is a concern for other countries as well as Spain, but also that it worsened there over the analysed period. This is probably due to the GFC and to rising energy prices. Other studies reveal that Eastern Europe has had the greatest deterioration in this regard in the EU (Buzar, 2007; Tirado and Ürge-Vorsatz, 2010), followed by Southern Europe (Healy and Clinch, 2002). In the case of Spain, the proportion experiencing difficulties paying their energy bills has almost doubled. A similar trend is visible among households claiming to be unable to maintain their home at an adequate temperature during the cold months (up from 8 per cent in 2007 to 11 per cent in 2014). The Spanish figures are the fourth worst in the EU on this indicator. According to Eurostat, the price of an electricity bill in an average home in Spain increased by 76 per cent between 2007 and 2014, in part due to the incorporation of costs associated with social and environmental policies. Likewise, the natural gas bill in an average Spanish home increased by 35 per cent over the same period. These data suggest that, in 2014, approximately 17 per cent of Spanish households experienced difficulties paying for electricity, gas or water.

In response to the serious economic problems faced by many families during the crisis, the Alliance against Energy Poverty (APE) and the Platform of

Mortgage Victims (PAH) were founded in Barcelona to 'pressure and denounce the public administrations for their immobility in relation to the extortionate charges of companies supplying energy and water' (APE, n.d.) and to 'modify mortgage legislation allowing the delivery of housing to banks to cancel all the debt' (PAH, 2013: 13). In Catalonia, in July 2015, Parliament unanimously approved the passing of the Popular Legislative Initiative promoted by PAH and APE. In general, the municipalities and the Autonomous Communities have to deal with this precariousness in accessing basic energy services. Thus, in January 2016, the government of the Community of Valencia dedicated €5.9 million to combating energy poverty, and Madrid dedicated €2 million at the end of 2015.

In spite of the difficulties in obtaining indicators that encapsulate sustainability in its most holistic form in an urban dwelling context, it can be concluded that the vulnerability of families since the GFC has increased due, among other things, to factors that are associated with housing in one way or another, among other factors. Both the difficulties of gaining access to a decent dwelling and the issue of energy poverty directly impact the well-being of Spanish families.

Barriers, drivers and challenges for sustainable communities in Spain

As mentioned above, Spain has no real tradition of concern for sustainability. In particular, environmental dimensions have been missing from local, regional and national agendas until very recently.

The 1998 Land Law adopted a philosophy of streamlining the urban planning process and stipulated that all land was available for development, except for areas of special ecological value. This differed from previous regulations, where only land necessary for satisfying housing needs could be designated for development. The rationale for the change was that making a large quantity of land available for development on the market would lead to a falls in prices. This did not happen.

Rather, this law led directly to the environmentally unsustainable development of coastal, mountain and metropolitan areas. This sprawling development, promoted by city councils and accepted or even encouraged by the Autonomous administrations, multiplied land occupation, increased the need for private transport and spoiled many natural areas. Moreover, the rapid urban development did not take into account the negative effects it might have on energy efficiency and passive energy use of buildings and urban space. The development was also economically unsustainable, since it has given rise to a vast residential stock with maintenance and urban services, management needs that are putting serious pressure on municipal coffers, particularly in areas where occupancy has not reached 100 per cent. This pressure is even greater in the case of sprawling cities, since maintaining sprawling urban developments is leading various municipalities throughout Spain towards bankruptcy. Finally, the development is socially unsustainable, because it has not met the housing needs of a significant proportion of the population with insufficient resources, forcing them to move to areas of high deprivation where affordable housing can be found, causing social segregation and compounding problems of social cohesion.

By contrast, the current legislation – the 2007 Land Law – argues that

> urban planning should respond to the requirements of sustainable development, minimizing the impact of the growth of cities and backing the regeneration of the existing city [as opposed to an economic development approach involving above all the creation of 'more' city] and seeking a compact rather than a sprawling or disorganized city model, as land is an economic resource, but also a natural, scarce, and non-renewable resource.
>
> (Official State Gazette, 2007: 23267)

This is clearly a reflection of concerns relating to the negative effects of real estate expansion with the compliance of urban planning principles.

In order to evaluate sustainable communities, it is important to emphasize that Spain presents different levels of achievement depending on the distinct characteristics of each Autonomous Community and each local government. Additionally, the specificities of each municipality together with their different points of departure make it difficult to carry out a general assessment of the barriers, drivers and challenges of sustainable communities.

This diversity makes it difficult to gather comparable and, at the same time, sufficiently homogeneous data. For example, it is hard to know the situation of Agenda 21 in Spain because of the lack of reliable data. Most of the municipalities that committed to initiating the Agenda 21 process remain in the development phase and few have gone on to produce concrete actions. Additionally, few have made a plan for monitoring the actions – a process that would entail the ongoing review of the approved measures.

Sustainable communities require agreement on long-term strategies between the different governance agents in the territory. This is particularly relevant for countries like Spain, where institutional thickness is very high. While policy transfer across 16 Autonomous Communities is high, collaboration remains very low (Aguado *et al.*, 2007). Local administrations, which are ultimately responsible for building sustainable communities, are extremely varied across Spain in terms of their size, amount of urban land, population density, and so on. This variability means that, in order to achieve more sustainable communities, the goals that are set have different degrees of scope.

Likewise, Aguado *et al.* (2007) reveal that public–private collaboration or partnerships in areas related to sustainability have also encountered difficulties. Public–private partnerships are not only a question of sharing financial burdens but also entail a strong willingness to cope together with adversity and define future strategies. However, the financial crisis has resulted in a significant reduction in public spending, with cutbacks in resources for goals with medium- or long-term results. Today, a significant proportion of local budgets is spent on solving emergencies caused by the crisis, such as evictions or unemployment, by providing, for instance, temporary shelter for evicted people or children's meals to guarantee a minimum standard of living. Private actors' involvement has mostly been limited to raising funds in reaction to the public purse's inability to cover all of the expenses related to combating the negative consequences of the crisis.

Partnerships were also generated between the government and the Network of Local Sustainable Development Networks (e.g. Portal Ecourbano, whose main objective is to disseminate projects, initiatives and tools that help build more sustainable cities; its ultimate goal is to reduce the impact of urban systems on the environment and improve the quality of life of inhabitants). However, it is worth emphasizing the role of BCNecologia, the Urban Ecology Agency of Barcelona, as a driver of initiatives like Ecourbano. This is a public consortium dedicated to rethinking cities in terms of sustainability: 'BCNecologia applies a systematic approach to redirect the management of cities towards a more sustainable model, providing solutions in mobility, energy, waste, urban planning, water, biodiversity, and social cohesion' (BCNecologia, 2012: 147). It has developed or participated in several projects not only in Spain but around the world, such as CAT-MED (Changing Mediterranean Metropolises around Time).

In general, Spanish city councils need to move beyond rhetoric and start to create strategies and programmes with concrete and measurable actions (Blanco and Subirats, 2008). In this regard, one of the most serious problems faced by Spanish society as a whole is the lack of enforcement. The Earth Programme, which was developed ahead of the 2015 elections by non-profit organizations such as Greenpeace, highlights the need to back non-speculative urbanism that is adapted to the people. In particular, it recommends the enforcement of regulations, the fight against speculative urban planning, and the fostering of the compact city.

The implementation of Local Agenda 21 in each area may be considered an indicator of the achievement of sustainable communities in Spain, despite the rather limited effects of its actions. But, here, many municipalities lose the holistic perspective and consider purely environmental dimensions without including economic or social aspects (Aguado *et al.*, 2007).

One important driver of sustainable communities is citizens' growing awareness of the need to develop measures to correct social, economic and environmental imbalances. Placing the citizen at the centre of decision-making and empowerment facilitates the pursuit of non-speculative goals dedicated to the sustainable improvement of the quality of life of citizens in general. The city of Barcelona is a good example in terms of awareness of the achievement of sustainable communities. It is worth highlighting, for example, the streamlining of land-based public transport or 'open government', which involves the participative online processing of all kinds of procedures. The main reason why the city was awarded the 'Capital of Innovation' prize in 2014 was 'for introducing the use of new technologies to bring the city closer to citizens' (European Commission, 2014). This took place through a project, 'Barcelona as a People City', launched in 2011, which aimed to introduce new technologies to foster the growth and welfare of citizens by providing, among other things, sustainable city growth initiatives on smart lighting, mobility and residual energy.

The 2016 study *Income Distribution, Economic Crisis and Redistributive Policies* (Goerlich Gisbert, 2016), funded by the BBVA Foundation and the Valencian Institute of Economic Research, examines the impact of the GFC on the tendencies of social stratification: at present, per capita and per household income have regressed to late twentieth-century levels. Not only has income fallen but its distribution has worsened since the start of the crisis, with inequality at a historic high. This increase in inequality, combined with a significant fall in income, has brought about a fall in living standards among the lower echelons of the population. Achieving a society that is less vulnerable to shocks would definitely reinforce the objectives associated with sustainable communities.

Environment, second homes and sustainability

Of the countries examined in this volume, Spain ranks third highest on the Environmental Protection Index (Table 1.1). However, certain areas which relate to excessive housing development (particularly tourist areas intended to accommodate second homes) have seen their environmental balance damaged in an irreversible way. Concentrated on the Mediterranean coast and the Balearic and Canary islands, the construction of second homes has followed a pattern of widespread, continuous and massive urbanization, which is modifying natural and social characteristics (Algualcil *et al.*, 2015). The burst of the credit bubble has led to significant growth in vacant dwellings located mainly in areas along the Mediterranean coast and other areas where there was an explosion in the construction of second homes. This has resulted not only in the destruction of the landscape but also the creation of 'ghost' housing estates with no residents that face difficulties on the property market.

The Population and Housing Census (INE, 2011) reveals that 29 per cent of the total housing stock in Spain is either used as second homes (15 per cent) or unused and vacant (14 per cent). Furthermore, the ECB's Household Finance and Consumption Survey (ECB, 2014) indicates that 36 per cent of Spanish households have a second residence in addition to their main home – the third highest figure in the EU after Cyprus (52 per cent) and Greece (38 per cent), and significantly above the levels found in the Netherlands (6 per cent), Austria (13 per cent) and Slovakia (15 per cent). This reveals standards in consumption, traditions, norms and habits that are very different from those in most of the European Union, as well as priorities in terms of family spending. The high level of second-home ownership in Spain is also linked to the rural homes that were abandoned during the internal migrations of the 1950s and 1960s. These subsequently became second homes for those who inherited them. When comparing with other European countries, particularly those that are wealthier than Spain, only 18 per cent of German families have a second home and only 25 per cent of French and Italian families (see Figure 3.3).

Foreign investment in the second-home market is another important factor, and sales to foreign investors have increased since the crash. At the beginning of 2009, foreign buyers represented just 5 per cent of the total, but by the third quarter of 2015 this figure had increased to almost 18 per cent (Observatory of Housing and

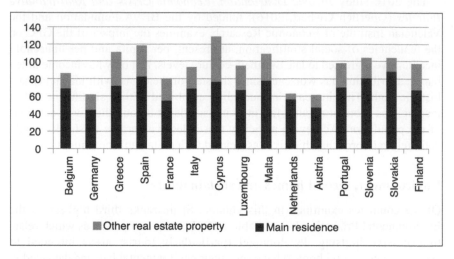

Figure 3.3 European households' real estate, 2010.

Source: ECB, 2014.

Note: Data for Greece, the Netherlands and Finland, 2009; for Spain, 2008.

Land, 2015). Foreign investment in real estate doubled in the five years from 2009 to 2014, reaching over €10 billion in 2015, which represented 18 per cent of total purchases (Observatory of Housing and Land, 2015). According to information from the Spanish Property Registrars (2016), 23 per cent of foreign home-buyers are from the United Kingdom, followed by France at 9 per cent, and Germany, Sweden and Belgium at 6 per cent each.

One way of understanding second homes is as an instrument for spreading the city into the countryside (Delgado Viñas, 2012). Housing construction in the period of real estate expansion found one of its strongest growth components in second homes (OS, 2006). It is precisely in the case of second homes where the boundary between consumption and investment uses in housing is at its vaguest. Exogenous negative shocks (such as the recent GFC) leave households with less income and they might find some relief from the sale of residential assets that do not serve as their main residence. This process has been widespread in Spain over recent years. The considerable increase in second-residence supply explains why the number of unsold vacant homes has had a considerable impact on areas where second homes proliferate.

In Spain, as in Ireland, the authorities directly or indirectly supported the housing development, with most local councils strongly in favour of additional construction for second-home owners (Paris, 2011). Spain has been subjected to a significant level of speculation, often facilitated by law and, at times, resulting in serious corruption. This can be related to historical difficulties with local public finances – there have been significant deficit problems since democratization of the city councils began. This situation often results in the rather lax (although often legal) granting of building licences, which can cause serious damage to the country's environmental balance (Delgado Viñas, 2012; Módenes and López, 2005).

Thus, the Spanish housing system faces a complex set of problems. First, there are the twin problems of increased incentives associated with real-estate speculation and rising levels of second-home consumption. As a consequence, the natural environment has been severely damaged in a sizeable area of Spain. Second, the growing number of vacant homes in second-home areas has led to widespread entry from foreign investors. While this is not negative in itself, in some places most of the investment has transformed towns and developments into gated communities, far from the ideal of sustainable communities (Haug *et al.*, 2007), where a mix of uses (i.e. residential, productive, green spaces), different population segments, interaction with the area and social cohesion is to be expected. It is interesting that large investors in the real-estate sector, such as PWC (2015), point out that the only investment opportunities in Spain for new housing promotions are found in highly attractive cities and on the coast, as there is a significant amount of vacant housing to absorb due to the mismatch between supply and demand. In addition, permanent residents in second-home areas might experience pressure on prices due to foreign investment, hindering their housing affordability: for instance, in 2015, the Balearic Islands were the most expensive Spanish region to buy a property in relation to local income, since there the average home costs 15.8 years of gross income (Sociedad de Tasación, 2015).

Finally, in Spain there is a strong correlation between neighbourhood density and the probability of owning a second home, and Módenes and López-Colás (2007) warn of the dangers associated with this. If politically correct urban planning in Spain prefers high-density land development, it ought to take into account the far greater relative increase in land area developed for second homes that this would entail. The higher the density of urban land, the greater the potential demand for second homes (Módenes and López-Colas, 2007).

Conclusions

Spain is characterized by a very high degree of both institutional and socio-economic diversity. There is significant variability in terms of the implementation and development of sustainable communities in urban environments. Nevertheless, there is also a series of general elements that characterize the current situation with regard to this issue.

The Spanish economy has been one of the most affected by the GFC and this has had a dramatic impact on the country's citizens. The high unemployment rates and precariousness in existing jobs, especially for young people, make it difficult for households not only to access housing but also to access basic needs such as electricity, water and gas. Energy poverty, for example, has considerably increased since the start of the crisis.

The predatory urban development of the period of expansion has had serious consequences, particularly in those areas where the construction boom reached especially high levels. Urban planning legislation in Spain constitutes one of the most serious problems when it comes to future changes in the economic situation and upswings in the real-estate sector. In order to make a change in urban planning matters, not only is a reconsideration of the function of urban planning needed, but also, and above all, the full implementation of existing responsibilities and regulations. In short, as well as reviewing principles, we need to comply with urban planning legislation and in the process move away from the speculative urban development that has taken place over recent years.

In Spain, the term 'sustainable community' is generally not used, allowing room for more generalist views often associated with an urban planning conception of the city (i.e. compact or urban sprawl) and focused especially on the form the city takes. The compact city is associated with greater sustainability and more balanced urban growth. To a large extent, the traditional Mediterranean compact city model favours the creation of sustainable communities. For this, measures are needed to promote urban regeneration and improve liveability in consolidated city centres where most places of employment are, rather than constructing new housing stock while hundreds of thousands of existing houses lie empty. Additionally, backing Mediterranean compact city forms over the dispersed city will help to reduce land consumption and mobility needs.

In Spain, the lack of a strong welfare state has paved the way for a social fabric based on networks of relatives and friends and the development of a significant role for civil society associations. This can be observed in the assistance of families in first-time access to housing and even in the guarantees provided for obtaining mortgages. Similarly, family networks have played an essential role in the rehousing of those who have been evicted from their primary housing. The empowerment of society in multiple areas of action due to the effects of the GFC can be seen in planning itself: from participative models of urban planning to unplanned changes in the use of urban spaces. Housing, as a central core of urban planning, emerges as a key element for the articulation of residential spaces according to the needs and demands of citizens.

Urban housing in sustainable communities should take on the role of a con-sumption good rather than an investment asset to allocate families' savings under speculative incentives. Spain must move away from the combination of the two that has taken place until now. The issue of vacant housing requires the develop-ment of a vacant housing census as well as plans to incentivize its use through social rent or similar measures, such as taxation. Moreover, municipal land assets owned by city councils should be used exclusively for subsidized rental and social housing. Property tax penalties for having vacant housing or property tax reduc-tions for energy efficient housing represent two potential measures for correcting imbalances and stimulating the sustainability of urban housing.

In summary, we are currently living through a period that offers opportunities to change the dominant paradigm, including in urban planning. The major blow from the GFC is intimately linked to the evolution of the construction sector and, consequently, to ways of city-making, of building, and of using urban planning. The financial sector and governments through public policies were not disconnected from the speculative bubble of the real-estate sector in the first decade of the twenty-first century. In general, there are two ways of approaching the analysis of the status of sustainable communities in Spain: on the one hand, understanding the crisis as an opportunity that requires a reconsideration of how we make cities, with much more participative and environmentally respectful standards; or, on the other hand, denying the exhaustion of the model and supporting its continuation and its capacity for economic growth.

References

Aguado, I. and Etxebarria, C. (2003) 'Medio ambiente y desarrollo sostenible en España', *Boletín Economico de ICE*, 2786: 21–30.

Aguado, I., Etxebarria, C. and Barrutia, J. M. (2007) *Indicadores de desarrollo humano sostenible: análisis comparativo de la experiencia española*. MPRA Paper No. 29019. Munich: University Library of Munich.

Algualcil, A., Alguacil, J., Arasanz, J., Fernández, G., Evangelista Paniagua, J. L., Olea, S., Renes, V. and Ayala (2015) *Housing in Spain in the Twenty-first Century. Executive Summary*. Madrid: Cáritas and Fundación Foessa.

Allen, J., Barlow, J., Leal, J. *et al*. (2004) *Housing and Welfare in Southern Europe*. Oxford: Blackwell.

Alliance against Energy Poverty (AEP) Website. Available at http://pobresaenergetica.es/es/que-es-ape/, accessed 7 August 2016.

Amore, K., Baker M. and Howden-Chapman, P. (2011) 'The ETHOS definition and classification of homelessness: an analysis', *European Journal of Homelessness*, 5(2): 19–37.

BCNecologia (Urban Ecology Agency of Barcelona) (2012) *Certificación del urbanismo ecosistémico*. Available at www.bcnecologia.net/sites/default/files/publicaciones/docs/certif_urb_ecosistemico_web.pdf, accessed 7 August 2016.

Blanco, I. and Subirats, J. (2008) 'Social exclusion, area effects and metropolitan governance: a comparative analysis of five large Spanish cities', *Urban Research and Practice*, 1(2): 130–148.

Boardman, B., (1991) *Fuel Poverty: From Cold Homes to Affordable Warmth*. London: Belhaven Press.

Burón Cuadrado, J. (2008) 'Una política de vivienda alternativa', *Ciudad y Territorio Estudios Territoriales*, 40(155): 9–40.

Buzar, S. (2007) *Energy Poverty in Eastern Europe: Hidden Geographies of Deprivation*. Aldershot: Ashgate.

Camarero, L. (2009) *La población rural de España de los desequilibrios a la sostenibilidad social*. Colección Estudios Sociales No. 27. Barcelona: Fundación la Caixa.

Delgado Viñas, C. (2008) 'Vivienda secundaria y turismo residencial como agentes de urbanización y segregación territorial en Cantabria', *Scripta Nova. Revista Electrónica de Geografía y Ciencias Sociales*, 12: 269.

Delgado Viñas, C. (2012) 'Secuelas territoriales de la burbuja inmobiliaria en las áreas protegidas litorales españolas', *Ciudad y Territorio: Estudios Territoriales*, 174: 615–638.

ECB (2014) *Eurosystem Household Finance and Consumption Survey*. Available at www.ecb.europa.eu/pub/economic-research/research-networks/html/researcher_hfcn.en.html, accessed 7 August 2016.

European Commission (2009) *European Project on Fuel Poverty and Energy Efficiency.* Available at https://ec.europa.eu/energy/intelligent/projects/en/projects/epee, accessed 7 July 2016.

European Commission (2014) *Barcelona, European Capital of Innovation.* Available at http://europa.eu/rapid/press-release_IP-14-239_es.htm, accessed 7 August 2016.

Eurostat (multiple years) *European Union Statistics on Income and Living Conditions.* Available at http://ec.europa.eu/eurostat/web/microdata/european-union-statistics-on-income-and-living-conditions, accessed 7 August 2016.

Gal, J. (2010) 'Exploring the extended family of Mediterranean welfare states, or: Did Beveridge and Bismarck take a Mediterranean cruise together?', in M. Ajzenstadt and J. Gal (eds) *Children, Gender and Families in Mediterranean Welfare States*, 77–101. Dordrecht: Springer.

General Council of the Judiciary (2014) *Statistics.* Available at www.poderjudicial.es/cgpj/es/Temas/Estadistica-Judicial/Analisis-estadistico/, accessed 7 July 2016.

Gordon, D., Adelman, L., Ashworth, K., Bradshaw, J., Levitas, R., Middleton, S., Pantazis, C., Patsios, D., Payne, S., Townsend, P. and Williams, J. (2000) *Poverty and Social Exclusion in Britain.* York: Joseph Rowntree Foundation.

Goerlich Gisbert, F. J. (2016) *Distribución de la renta, crisis económica y políticas redistributivas.* Bilbao: Fundación BBVA.

Haug, B., Dann, G. M. S. and Mehmetoglu, M. (2007) 'Little Norway in Spain: from tourism to migration', *Annals of Tourism Research*, 34(1): 202–222.

Healy, J. D. and Clinch, P. (2002) *Fuel Poverty in Europe: Across-Country Analysis Using a New Composite Measure.* Environmental Studies Research Series. Dublin: University College Dublin.

Hoekstra, J., Heras, I. and Etxezarreta, A. (2010) 'Recent changes in Spanish housing policies: subsidized owner-occupancy dwellings as a new tenure sector?', *Journal of Housing and the Built Environment*, 25(1): 125–138.

Indovina, F. (2007) *La ciudad de baja densidad: lógicas, gestión y contención.* Barcelona: Observatorio Territorial de la Dirección de Estudios de la Diputación de Barcelona.

Instituto Nacional de Estadística (multiple years) *Censos de Población y Viviendas.* Madrid: INE.

Instituto Nacional de Estadística (2014–2015) *Encuesta de Población Activa (EPA).* Madrid: INE.

Jurado Guerrero, T. (2004) 'La vivienda como determinante de la formación familiar en España desde una perspectiva comparada', *Revista Española de Investigaciones Sociológicas*, 103(3): 113–157.

Local Agenda 21 (LA21) (2008) *Green Book of Urban Environment.* Available at www.bcnecologia.net/en/projects/green-book-urban-environment-volume-i-and-ii, accessed 7 July 2016.

Ministry of Environment (2009) *Consumo de energía por hogar y Emisiones de CO2 del sector residencial.* Available at www.magrama.gob.es/es/calidad-y-evaluacion-ambiental/temas/informacion-ambiental-indicadores-ambientales/2_12hogares_tcm7-2143.pdf, accessed 7 July 2016.

Ministry of Environment (2011) *EESUL Estrategia española para la sostenibilidad urbana y local.* Available at www.magrama.gob.es/es/calidad-y-evaluacion-ambiental/temas/medio-ambiente-urbano/EESUL-290311-web_tcm7-177531.pdf, accessed 7 July 2016.

Ministry of Public Works (2010) *Libro Blanco de la Sostenibilidad en el Planeamiento Urbanístico Español.* Available at www.fomento.gob.es/MFOM/LANG_CASTELLANO/

DIRECCIONES_GENERALES/ARQ_VIVIENDA/SUELO_Y_POLITICAS/ ESTUDIOS/Libro_blanco/, accessed 7 August 2016.

Ministry of Public Works (2013) *Informe sobre el stock de vivienda nueva 2013*. Madrid: Ministry of Public Works.

Módenes, J. A., Fernández-Carro, C. and López-Colás J. (2013) 'La formación de hogares y la tenencia de vivienda de los jóvenes en la reconfiguración de los sistemas residenciales europeos', *Scripta Nova. Revista Electrónica de Geografía y Ciencias Sociales*, 17(460). Available at www.ub.edu/geocrit/sn/sn-460.htm, accessed 7 July 2016.

Módenes, J. A. and López, J. (2005) 'Expansión territorial de la residencia secundaria y ciudad compacta en España: ¿elementos de un mismo sistema?', *Papers de demografía*, Working Paper 274.

Módenes, J. A. and López-Colás, J. (2007) 'Second homes and compact cities in Spain: two elements of the same system?', *Tijdschrift voor Economische en Sociale Geografie*, 98(3): 325–335.

Moliní, F. and Salgado, M. (2010) 'Superficie artificial y viviendas unifamiliares en España, dentro del debate entre ciudad compacta y dispersa', *Boletín de la Asociación de Geógrafos Españoles*, 54: 125–147.

Navarro, V. (2004) *El estado del bienestar en España*. Barcelona: Editorial Tecnos.

Navarro, V., Torres, J. and Garzón, A. (2012) *Hay alternativas. Propuestas para crear empleo y bienestar social en España*. Madrid: Sequitur.

Observatory of Housing and Land (2015) *Ministerio de Fomento, Dirección General de Arquitectura, Vivienda y Suelo. Boletín núm. 15. Tercer trimestre*. Available at http://publicacionesoficiales.boe.es/detail.php?id=000716115-0001, accessed 7 July 2016.

Observatory of Sustainability (OS) (2006) *Sostenibilidad en España*. Available at www.urv.cat/media/upload/arxius/W-Catedra_DOW_URV/Informes%20VIP/ose_-_informe_2006.pdf, accessed 7 August 2016.

Observatory of Sustainability (OS) (2014) *Sostenibilidad en España*. Available at www.observatoriosostenibilidad.com/documentos/SOS16_v21.pdf, accessed 7 August 2016.

Observatory of Sustainability (OS) (2016) *Sostenibilidad en España*. Available at www.urv.cat/media/upload/arxius/W-Catedra_DOW_URV/Informes%20VIP/ose_-_informe_2006.pdf, accessed 7 July 2016.

Odyssee-Mure Energy Efficiency Database (multiple years). Available at www.indicators.odyssee-mure.eu/energy-efficiency-database.html, accessed 7 July 2016.

Official State Gazette (2007) *LEY 8/2007, de 28 de mayo, de suelo*. Available at www.boe.es/boe/dias/2007/05/29/pdfs/A23266-23284.pdf, accessed 7 July 2016.

Omdbusman (2013) *Informe sobre Viviendas Protegidas Vacías*. Available at www.defensordelpueblo.es/es/Documentacion/Publicaciones/monografico/contenido_1363855813805.html, accessed 7 July 2016.

Pareja-Eastaway, M. (2007) 'Residential opportunities and emancipation strategies in Spain', *Architecture, City and Environment*, 2(5): 453–459.

Pareja-Eastaway, M. and Sánchez-Martínez, M.-T. (2014) 'Spain', in T. Crook and P. A. Kemp (eds) *Private Rental Housing*, pp. 71–98. Cheltenham: Edward Elgar.

Pareja-Eastaway, M. and Sánchez-Martínez, M.-T. (2015) *Is the Private Rented Market Filling the Role of Social Housing in Spain?* London: ENHR Working Group, LSE.

Pareja-Eastaway, M. and Sánchez-Martínez, M.-T. (2016) 'Have the edges of homeownership in Spain proved to be resilient after the Global Financial Crisis?', *International Journal of Housing Policy*, special issue: *The Edges of Homeownership*. Available at www.tandfonline.com/doi/abs/10.1080/14616718.2016.1185275?journalCode=reuj20, accessed 7 July 2016.

Paris, C. (2011) *Affluence, Mobility and Second Home Ownership*. Abingdon: Routledge.

Piedrafita, S., Steinberg, F. and Torreblanca, J. I. (eds) (2006) *20 Years of Spain in the European Union (1986–2006)*. Madrid : Elcano Royal Institute.

Platform of Mortgage Victims (PAH) (2013) *Emergencia habitacional en el Estado Español*. Available at http://afectadosporlahipoteca.com/wp-content/uploads/2013/12/2013-Emergencia-Habitacional_Estado_Espanyoldef.pdf, accessed 7 August 2016.

Price Whaterhouse Coopers (PWC) (2015) *Residential Dynamics in Spain*. Available at www.pwc.es/en/publicaciones/construccion-inmobiliario/assets/trebol-news-abril-2015-en.pdf, accessed 7 July 2016.

Romero, J. (2010) 'Construcción residencial y Gobierno del Territorio en España. De la burbuja especulativa a la recesión. Causas y consecuencias', *Cuadernos Geográficos de la Universidad de Granada*, 47(2): 17–46.

Sánchez-Martínez, M.-T. (2008) 'The Spanish financial system: facing up to the real estate crisis and credit crunch', *International Journal of Housing Policy*, 8(2): 181–196.

Sociedad de Tasación (2015) *Spanish Property Insight*. Available at www.spanish propertyinsight.com/about/, accessed 7 July 2016.

Spanish Property Registrars (2016) *Boletín Estadístico Registral*. Available at www.registradores.org/portal-estadistico-registral/boletin-estadistico-registral/, accessed 7 Augist 2016.

Tirado Herrero, S. and Ürge-Vorsatz, D. (2010) *Fuel Poverty in Hungary: A First Assessment*. Budapest: Centre for Climate Change and Sustainable Energy Policy.

Thomson, H. and Snell, C. (2013) 'Quantifying the prevalence of fuel poverty across the European Union', *Energy Policy*, 52(C): 563–572.

Valenzuela, M. (2008) *Progresos hacía el modelo urbano español más sostenible*. Available at www.uam.es/gruposinv/urbytur/documentos/01.pdf, accessed 7 July 2016.

Verdaguer Viana-Cárdenas, C. (2011) *Informe de situación de las principales actuaciones e iniciativas en materia de sostenibilidad urbana en España*. Grupo GEA 21, Monografía (Informe Técnico). Madrid: E.T.S. Arquitectura (UPM).

Vinuesa Angulo, J. (2008) 'La vivienda vacía en España: un despilfarro social y territorial insostenible', *Scripta Nova. Revista Electrónica de Geografía y Ciencias Sociales*. Available at www.ub.edu/geocrit/-xcol/74.htm, accessed 7 July 2016.

4 Sweden

Inga-Britt Werner

Introduction

This chapter assesses the sustainability of housing and communities in Sweden. Sweden has a relatively high GDP per capita, among the five richest of countries examined in this volume, and a long history of being a welfare state with stable political majorities. According to the IMF, the Swedish national economy did not suffer to the same extent as many other EU countries from the global crisis of 2008. This was due to efficient stabilisation measures for regulating banks' lending plus the acceptance of an occasional modest national budget deficit (IMF, 2011). GDP has been positive since 2010, although annual growth is within the range of 1 to 3 per cent (SCB, 2015a). The total burden of taxation was relatively high at 44 per cent of GDP in 2012, compared with 39 per cent on average for EU28 (Eurostat, 2015a). These taxes support a welfare regime that, although political majorities change, can be characterised as social democratic according to Esping-Andersen's typology of welfare regimes (Esping-Andersen, 1990). Policies of public support to households in need still hold in periods of liberal–conservative government, the latest of which was between 2006 and 2014. Since 2014, the Swedish government has been an alliance of Social Democrats and Greens, 'the Environment Party'.

Sweden is a unitary state, but substantial responsibilities are devolved to 290 municipalities. The latter manage a wide range of mandatory tasks of community development and welfare. They are most important actors in the fields of housing and sustainability, responsible for land use and the built environment together with social welfare. Municipalities handle planning, building permits, household and industrial waste, water and sewage, protection of the local natural environment, social benefits to persons in need, education up to university level, day care of children, care of the elderly and public transport (the latter in cooperation with county municipalities). Resources for all this come partly from taxes transferred by the government and partly from a flat rate tax of 30–32 per cent on residents' personal income. The municipal tax rate is decided by democratically elected municipal councils.

The population of Sweden is rising significantly. According to Eurostat, during 2014 the Swedish population growth rate was the highest of all the EU28 countries

(Eurostat, 2015b). According to Statistics Sweden, the annual net increase has varied between 0.7 and 0.9 per cent since 2005. During 2015, the preliminary net increase rose to 1.1 per cent (SCB, 2015b). In 2014, 75 per cent of population growth was due to net immigration and 25 per cent to natural growth (SCB, 2015c). The increasing inflow of refugees from Middle Eastern and African war zones is a significant component in the Swedish population growth. As a result of this growing population, there is a need for more affordable housing, especially in the larger cities.

The geographic conditions of Sweden are important for a discussion on sustainability and comparisons in a European context. Sweden is situated at the northern periphery of the EU, and the country's northernmost part is crossed by the Arctic Circle. Climatic conditions range from temperate to sub-arctic. The relatively cold climate makes energy consumption for heating an important issue. Energy efficiency of buildings has been part of Swedish policies for sustainability since the first oil crisis in 1973. Heating of the built environment makes up a substantial part of the national energy consumption – 21.4 per cent in 2013 – on a par with the total consumption of energy for domestic transport that year (Energimyndigheten, 2015).

The housing system in Sweden

In this section I present a brief overview of the housing system in Sweden. It is generally referred to as entailing a unitary rental model (Kemeny, 1995; Kemeny *et al.*, 2005) . This was particularly true from 1950 to around 1990. The national government then issued detailed regulations on how, where and how much to build. In return, public housing companies owned by the municipalities received large governmental subsidies in the form of guaranteed low interest rate on borrowed capital to fulfil national production goals. Rents were strictly controlled, also in the private renting sector. This system has changed profoundly since the beginning of 1990s, as subsidies have been abandoned and rent controls relaxed. For a more comprehensive discussion of these recent changes, see Hedin *et al.* (2012), Christophers (2013), Söderholm (2013) and Andersson and Magnusson Turner (2014).

The Swedish government has no specific department of housing. Housing policy is determined under the auspices of the Ministry of Enterprise, Energy and Communications. There is one national organisation dedicated to housing issues: Boverket – the National Board for Housing, Building and Planning. The main tasks of this institution are to analyse national and regional housing markets, issue building regulations and supervise physical planning. Boverket supports municipalities in their role as the main actors responsible for housing. This responsibility includes the supply of high-quality, sustainable housing to meet national political goals. The municipalities' most important tool for initiating new construction of housing is planning. By issuing legally binding, detailed land development plans that are conditional for building permits, each municipality has the exclusive right to supply land for construction purposes. Since 2014 the municipalities have been required to issue policy guidelines for achieving a balanced local housing market and fulfilling national housing goals. There is no requirement to express the guidelines in quantitative terms per annum. Boverket carries out an annual survey with all Swedish municipalities on assessed housing need and calculates figures on the national and regional levels. In 2015 it declared that 71,000 new units should be built annually between 2015 and 2020 (Boverket, 2015a) – that is, double the number that were built in that year.

In 2014, rented dwellings represented 38.2 per cent of Swedish housing stock (SCB, 2016). As a legacy from the period of housing production based on state subsidies, most Swedish municipalities own one or more housing companies for renting out dwellings. In 2012, these public rental companies held 46 per cent of Swedish rental flats (Boverket, 2014), which makes them a substantial part of the Swedish housing market. The private rental sector works under the same laws as the public rental sector regarding negotiation of rent levels and allocation of dwellings. The Swedish system for setting rents is complicated, but in short it is based upon collective bargaining between local organisations of landlords and tenants. There are no by-laws on affordability of rents but tenants' associations are tasked with this issue. Rent levels in new housing are based on production costs, including the price of land, and to some extent on companies' evaluation of housing quality. Prospective individual tenants cannot outbid sitting tenants to obtain a rental contract. Regional rent and tenancies' tribunals judge whether negotiated rents are fair, compared to other dwellings of the same type, standard and location. All rental dwellings, including those owned by private enterprises, are allocated through queuing. In larger municipalities a municipal housing authority administrates the queue. Municipalities can decide on criteria for priority, such as being a single parent, a family living in crowded conditions or a recent immigrant (Christophers, 2013).

This housing system makes private actors in the rental market deem production of new rental housing less profitable, as they are not free to set rents the market would be willing to pay. Since 2011, public rental companies are bound by law to operate on a commercial basis with the aim of yielding returns to their owners, the municipalities. To achieve this, many companies are selling part of their housing stock to gain capital to refurbish their remaining stock. Their capacity for new production is limited. As a result, new production of rental housing is low. To remedy this, in 2014 the government reintroduced subsidies for new construction of small rental dwellings at a rent that does not exceed a given maximum. The aim of this policy is to facilitate young people to establish their first homes.

There is formally no 'social housing' sector in the Swedish housing system, in the sense that certain dwellings are allocated only to low-income households. As a means to enhance the affordability of housing the welfare system offers housing benefits, on criteria of total household income, household composition and current housing cost and dwelling space. The housing benefits target only households with children still at school, young people (between 19 and 29 years of age) and pensioners. In 2014 the absolute maximum of housing allowance per household per month was €560, but only households with three or more children and an annual income of less than €12,595 are eligible for that amount (Försäkringskassan, 2015). Even though there is no social housing sector in Sweden, neighbourhoods where income and education levels are low do exist. In many of these neighbourhoods a majority of the properties are rental and constructed between 1960 and 1970 by municipal housing companies. The large-scale housing blocks from this period quickly became less attractive to Swedish middle-class households, as ownership of one-family homes became increasingly available to them. Vacancies appeared in the rental stock and households with low income or on social welfare moved in. Over time, such neighbourhoods gained a reputation as less attractive and unsafe, which led to further segregation.

Beside rental tenure, in 2014, 22.3 per cent of Swedish dwellings were in tenant ownership tenure (SCB, 2016). Tenant ownership refers to cooperative ownership, where tenants are shareholders in an association owning the property. In Swedish the tenure is called '*bostadsrätt*', literally 'the right to occupy a dwelling'. Shares of the association, that is the 'right of occupancy' for a dwelling, are sold at market prices and can be mortgaged. This form of tenure has dominated the production of multi-family housing in Sweden since the early 2000s. Private developers have stepped in to produce housing for a market and to minimise risk; new construction projects do not start until the developer has listed a secure majority of interested buyers for the planned units. Individual ownership of property is the dominant tenure for one- and two-family dwellings and in 2014 owner-occupied dwellings made up 39.5 per cent of Swedish housing stock (SCB, 2016).

Sustainable communities and housing in Sweden

Definitions and policies for sustainable development in Sweden

The sustainable development of a society is a goal that is highly dependent upon political decisions. Thus, a definition adopted at the highest political level, the government, is of interest. In 2003, the government of Sweden presented a strategy for a sustainable society, including housing and communities:

> A sustainable society is a society where economic growth, social welfare and cohesion are linked to high environmental quality. This society meets its present needs without compromising the ability of future generations to meet their own needs. It is a society embracing democratic values.
>
> (Regeringen, 2003: 7; my translation)

Securing future generations' needs and linking three interdependent aspects of sustainability – economy, social cohesion and environmental quality – clearly echoes the Brundtland Commission's well-known definition of sustainable development (WCED, 1987). In 2006 the Swedish government's revised strategy stated that the overarching goal is to achieve economic growth under sustainable conditions. Regarding housing and communities, the same document states that sustainable communities should offer shops near housing areas, access to waste recycling stations, sustainable design of residential buildings and energy systems, plus access to sustainable transport (Regeringen, 2006).

Political goals are operationalised by policies, laws and regulations. Economic and social sustainability are in many ways less concrete and harder to regulate than environmental sustainability. Economic policy and the welfare regime are the main tools for implementing these goals. Regarding environmental sustainability, the Swedish regulatory framework focuses on energy consumption, the treatment of water and sewage, reducing harmful substances from industrial production and waste, conservation of the natural environment and biodiversity. Regulations stating maximum values of consumption and emissions are combined with taxes on consumption of, for example, electricity and fuel for vehicles.

Sustainable housing and communities indicators

Sweden has the lowest proportion of inhabitants in densely populated areas in the EU (see Table 1.1). Population density is also low by European standards, at 23.9 inhabitants/square kilometre (SCB, 2015d). The distribution of the Swedish urban population is characterised by concentration in the four largest cities, which are not very large as they have between 200,000 and 900,000 inhabitants. In addition, only nine more municipalities reach a population between 100,000 and 200,000 inhabitants. Sweden has a total of 290 municipalities, of which 95 per cent have populations below 100,000 inhabitants (Johansson and Haas, forthcoming). The average household size is among the lowest in Europe, similar to that found in other Scandinavian countries and Germany. These characteristics of Swedish population density are important when considering the development of sustainable communities as small cities face difficulties regarding the development of public transport and other collective services. The high number of single-person households puts pressure on the housing market, as the demand for small, affordable flats is very high. The rising proportion of older people is a growing concern in many countries, including Sweden. However, sizeable immigration combined with family-friendly policies has helped to reduce the challenge presented by an aging population. The proportion of the population over 65 years has risen steadily over recent decades so that in 2014 it was almost 20 per cent, but the proportion of the population aged 25–64 years has been fairly stable since the 1970s at approximately 50 per cent (SCB, 2015e).

The Swedish score on housing quality is high, second only to that of Norway (see Table 1.2). In addition to high performance on the aspects measured in that index, newer Swedish housing stock ranks highly with respect to other aspects of housing quality; limited energy consumption and 'design for all'. The latter is a concept describing built environments' adaptation for use by everyone, including people with disabilities. Increasing demands in these respects have obliged builders to achieve high levels of quality since the 1990s, although housing from this period represents a small proportion of the total housing stock (see Figure 4.1).

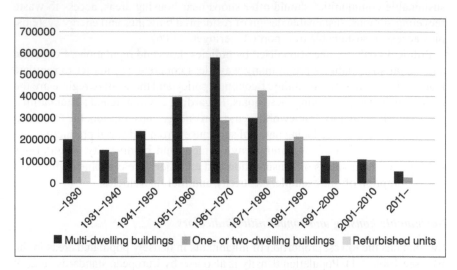

Figure 4.1 Swedish housing stock plus number of refurbished units, per year of construction.

Source: Collated by the author using data from SCB (2015f, 2015g).

Note: The number of refurbished units in each age category is shown in light grey bars. As the data on refurbished units cover only the period 1989–2008, refurbishment of the oldest housing stock is not fully represented in the diagram. A considerable amount of modernisation of old housing was carried out from the late 1970s onwards.

The Swedish housing stock is characterised by a large proportion of housing units in multi-family buildings constructed between 1950 and 1980. The construction of one- or two-family dwellings peaked between 1960 and 1980, but a substantial proportion of units were constructed before 1930. The many multi-dwelling buildings built between the 1950s and 1970s, more than a million units, are in need of renovation and modernisation. Figure 4.1 also illustrates the number of units that were refurbished between 1989 and 2008. The abrupt decline since 2008 is not due to the financial crisis, but to the abandonment of state subsidies for modernisation as well as a lack of data. Sveriges Byggindustrier – the Swedish Construction Federation – estimated that in 2013 €32,000 million was required for investment in modernisation. This is was 60 per cent of the total amount invested in new construction of housing in 2012 (Sveriges Byggindustrier, 2014). These figures exclude costs for improving energy performance, which means the need for investment is even greater, considering goals of energy efficiency.

Swedish building regulations on energy efficiency have become much more demanding over time, as shown in Table 4.1. The maximum values allowed for energy consumption for heating and hot water relate to the middle part of Sweden, where Stockholm is situated (Boverket, 2006, 2011, 2015b). For comparison, the table shows the average energy consumption for heating of housing (Energimyndigheten, 2014a, 2014b).

Table 4.1 Energy efficiency requirements for new housing 2006–2015, maximum kwh/square metre dwelling area/year

| | 2013 | 2006 | | 2011 | | 2015 | |
| | Average energy consumption – heating | Regulation, max consumption | | Regulation, max consumption | | Regulation, max consumption | |
	All energy sources kwh/sq m	Electricity kwh/sq m	Other sources kwh/sq m	Electricity* kwh /sqm	Other sources kwh/sq m	Electricity* kwh/sq m	Other sources kwh/sq m
One- and two-dwelling buildings	110**	75	110	55	90	55	90
Multi-dwelling buildings	139***	–	110	55	90	50	80

Sources: Boverket, 2006, 2011, 2015b; Energimyndigheten, 2014a, 2014b.

Notes:
* Includes use of electricity for heat pumps.
** One- and two-family buildings for permanent occupancy.
*** Multi-family buildings include heated floor space for common use (e.g. storage and garages).

Surprisingly, energy consumption per square metre is substantially higher in multi-family buildings than in one- and two-family houses. However, it includes energy consumption in the common areas of such buildings. The heating systems of Swedish multi-family dwellings are generally designed to operate for the whole building, giving little individual control over indoor climate. Heating costs in multi-family dwellings are included in the rent or in the fee to the ownership association, which makes energy conservation less appealing for individual households. Home owners benefit economically from energy conservation, as they pay the bill directly to the energy producers. They are also able to control their heating and indoor climate (Nairn *et al.*, 2010; Engvall *et al.*, 2014). One Swedish study states that individual home owners' indoor temperatures are around 2 degrees Celsius below those of residents in multi-family dwellings (Lindén *et al.*, 2006). This indicates that targeting behaviour by giving households' control and incentives to conserve energy would be a fruitful approach.

Table 4.1 reveals that regulations are especially demanding when heating by electricity. For one- and two-family dwellings, the 2015 requirement for energy consumption is less than half the 2013 actual average consumption for those heating their homes by electricity. Multi-family buildings must meet even more exacting standards. The particular focus on saving electricity has a link to Swedish electricity production, where 41 per cent is produced by water power (Energimyndigheten, 2015), as the northern part of the country is mountainous with many rivers. The conditions for CO_2-lean electricity production are thus favourable. However, Sweden still produces 43 per cent of its electricity via nuclear power (Energimyndigheten, 2015), which is of debatable sustainability.

Barriers, drivers and challenges for sustainable communities in Sweden

As in all societies, path dependency is a barrier against profound change in Swedish housing and communities. Most of the housing stock is already built and energy and transport systems are constructed under certain technical and economic conditions. To change such conditions will take time and political strength. The backlog of maintenance and modernisation of the large housing stock from the 1960s and 1970s is one such example, and a relatively high car dependency for personal transport is another. Economic considerations can form strong barriers as well as acting as drivers. Regarding Swedish nuclear plants, there has been a long battle between political parties that are totally against nuclear power, highlighting its risks, and parties in favour of its relatively cheap and local production of electricity, disregarding the risks. A series of proposals to close the plants has been put forward, but none of these has passed through parliament. So far, the nuclear plants are still working, but investment in new plants is not allowed. The older plants are now in need of renovation and production costs are rising. The economy might serve as a driver for change into other technologies. A case of the economy as a barrier is when green groups advocate a higher carbon dioxide tax on the Swedish industrial and transport sectors, but expected effects on Swedish competitiveness in the global market place are holding back the majority.

A growing awareness of the risks incurred by climate change is, on both global and local levels, a strong political driver for achieving more sustainable communities and housing. The present Swedish government points out the importance of measures and policies to counteract climate change. Its strategy for participation in COP21, the 2015 Paris Climate Conference, states that Sweden will act as a model country and will strengthen its domestic policies against climate change (Regeringen, 2015). The larger Swedish municipalities compete with each other to be at the forefront of technical and social development towards a more sustainable society, as this symbolic value is seen as a means to attract people. As can be seen from Table 1.1, Sweden has a high ranking on the international Environmental Protection Index. The aspects of the index that are linked to housing – climate and energy, health impacts, air quality, water and sanitation and water resources – are especially high (YCELP and CIESIN, 2014). Swedish policies regarding energy consumption are effective and broadly accepted. A number of targeted subsidies have led Swedish households, municipalities and property companies to invest in energy-saving technology (heat pumps, windmills, photovoltaic panels and district heating). The urban environment is relatively healthy as a result of urban planning regulations, although traffic noise and emissions are problematic close to traffic arteries.

The pending challenge now is an overheated housing market where prices are rising steadily. There are notable housing shortages in a majority of Swedish municipalities, and this problem is especially severe in the three largest cities. Housing shortages are worst in the metropolitan regions (Stockholm, Gothenburg and Malmö). The most significant shortage concerns affordable dwellings for young households and low-income groups, but medium-income groups are also affected. Housing production has increased slowly but exceeded 40,000 units per year only once between the early 1990s and 2014 (see Figure 4.2). The construction industry complains that the planning process is too slow and that there are too many regulations over energy efficiency. The supply of land for housing is limited as municipal urban planning takes a long time. Many actors are involved, with the noble intention of enhancing democracy in the planning process. This works to a certain extent, but in many cases local stakeholders obstruct the construction of new housing, campaigning only in their own interests (the NIMBY phenomenon). In addition, there is a lack of skilled construction workers, further increasing construction costs as builders are forced to pay skilled staff higher wages.

The present very high inflation in Swedish house prices is a serious concern for housing affordability. The tax system offers personal tax deductions for paid capital costs, which makes households more prone to take on large amounts of debt, fuelling the inflation further. The price boom is reinforced by historically low rates of interest and interest-only mortgages. In 2014, on average Swedish households were indebted up to a level of 175 per cent of their annual net household income, but for households mortgaged for buying a dwelling the

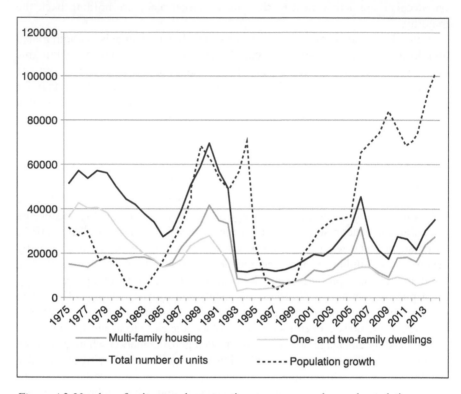

Figure 4.2 Number of units started construction per annum and annual population
    growth, 1975–2014.

Source: Collated by the author using data from SCB (2015b, 2015h).

proportion of debt to annual net income is 315 per cent (Sveriges Riksbank, 2015).
In addition, Swedish households were more indebted in 2013 than during the
GFC, which has caused some concern (Sveriges Riksbank, 2014). A 22 per cent
capital gains tax from sold property or shares of tenant ownership associations
discourages people from moving into dwellings that would better suit their needs,
keeping residential mobility low. This is a threatening home-spun crisis
(Holmqvist and Magnusson Turner, 2014).

A growing population, especially in the largest cities, together with a low rate
of housing construction, increases the risk of overcrowding. The proportion of
Swedish overcrowded households is relatively low in EU terms, but it is on
the rise, and it is already higher than in many Western European countries
(see Table 1.2). Among some groups, for example non-European immigrants, the
proportion of households in crowded conditions is much higher than the national
average, with estimates suggesting rates up to 40 per cent. Again this problem is
most notable in the largest cities (Boverket, 2015c; Holmkvist and Magnusson
Turner, 2014). The ongoing transformation from rental tenure into different forms

of ownership in the housing stock has resulted in a decline in the number of dwellings available to rent. Low-income households cannot compete for dwellings on the ownership markets and have to solve their housing needs by crowding into small flats or sublet rooms within a dwelling. At the same time, despite the housing deficit, there is some overconsumption of residential space in Sweden. According to Statistics Sweden, a household is considered to have a high standard of dwelling space if it has at its disposal more than one room per person in the household, not counting kitchen and living room. The proportion of households with this standard is substantial – 44 per cent in 2014 – and has been growing steadily since 2000. The percentage of overcrowded households has also risen over the same period (SCB, 2015i). Former housing policies resulted in significant construction of dwellings for families with children. The housing stock is now ill suited for a decline in household size (Söderholm, 2013). Low levels of residential mobility due to high transaction costs for moving house do not help. There are increasing inequalities regarding residential space consumption, security of tenure and housing costs, mainly between income groups but also between age groups and between residents of metropolitan areas and those living in smaller towns. Crowded conditions are common among low-income groups in the large cities and a high standard of dwelling space is most common among older households.

According to Boverket (2009), Sweden had among the highest proportion of housing expenditure to income in Europe in 2007, second only to Denmark. The proportion of housing costs to household net income differs by tenure. There is a growing gap between renters and owner households. In 2013, renting households had an average housing expenditure of 28 per cent of net income, whereas for home owners it was just 16 per cent (SCB, 2015j). This gap may be due to several factors: the tax system is favourable towards ownership; the interest rate on housing loans is historically low; and renting households tend to be young, old or recent immigrants. All of these factors are linked to lower incomes. Although Swedish house prices are high and rising, house owners' proportion of housing expenditure to income decreased from 19.5 per cent in 2004 to 15.7 per cent in 2013 (SCB, 2015j). Considering that the average annual housing expenditure is higher among owners than renters (SCB, 2015j), the higher proportion of housing expenditure to income among renters means that renters have relatively low incomes. This is in line with an observed development where low-income groups are filtered into rental flats, especially those of public rental companies (Andersson and Bråmå, 2004; Andersson et al., 2010; Hedin et al., 2012; Andersson and Magnusson Turner, 2014). Housing affordability and social segregation are now challenges for social sustainability.

Social sustainability via the right to buy? Case studies from Stockholm

The background to this discussion is rising home ownership in Sweden, both individual and cooperative, and the challenges this development brings. A decline in the number of rental units is due both to the policy of the recent liberal government to sell municipally owned properties to tenants' associations and to a lack of new construction for renting. Ongoing research on a municipal programme to sell public rental stock to tenants' associations in Stockholm offers concrete examples of the effects on social sustainability at the neighbourhood level. In 2008, Stockholm City, through its public rental housing companies, owned about 26 per cent of the total stock of multi-family dwellings in the municipality. A moderately neoliberal political majority in City Hall, elected in 2006, decided to sell a substantial part of this stock to tenants and private actors in the rental market. The aim of this policy of selling to tenants was to avoid neighbourhood segregation of public rental housing and to achieve social stability through social mix. The policy specifically targeted suburbs with lower than average incomes. The hypothesis was that ownership would promote responsibility for the shared residential environment, social cohesion and social sustainability.

Social sustainability is a complex concept. Dempsey *et al.* (2011) discuss its connections to environmental sustainability and related concepts such as social cohesion and social inclusion at the neighbourhood level. They conclude that social equity and a sustainable community are core aspects of urban social sustainability. According to these authors, social equity relates to the fairness of distribution of common resources and equality of conditions, measured in terms of, for example, access to education and public services. A sustainable community, in a broad sense, refers to a continuously viable and functioning community or neighbourhood (Dempsey *et al.*, 2011). Neighbourhood effects by social mix are a widely researched topic. It is assumed that social mix implemented in deprived neighbourhoods will improve residents' life chances through mechanisms such as peer influence, access to information on job opportunities and more effective social control, although causality is difficult to prove. (See, for example, Andersson *et al.*, 2007; Galster 2001, 2007, 2012; Hedman and Galster, 2013.)

The Stockholm study referred to here is a comparative case study. Two 'treated' cases –defined as neighbourhoods where a substantial part of the housing stock was converted to tenant ownership – were compared with two 'untreated' cases – neighbourhoods with properties in rental tenure – and, regarding socio-demographic changes and house prices, with the general development in the Stockholm municipality. The studied neighbourhoods, treated and untreated, were selected using the following criteria: a history as a deprived neighbourhood plus similarity in the age of buildings and population size. In the treated neighbourhoods, the tenure change was no earlier than 2005, and prior to the change housing stock had been owned by public rental companies. The selected cases are situated south of the Stockholm central business district. Each has a population of around 5,000 inhabitants. Mixed methods are employed, including qualitative interviews with key actors, documentary research and observation, quantitative surveys and analyses of existing databases. Data regarding the untreated neighbourhoods are still incomplete, so the discussion is mainly based upon findings from the treated neighbourhoods.

There is some evidence of gentrification in the two treated areas as the demographic profile of both changed immediately after the tenure change.

Employment, income and education levels all rose in these neighbourhoods. By contrast, in the comparison areas, income levels fell, unemployment rates rose and the proportion of immigrants increased during the same period (Werner and Annadotter, 2013). To buy, you must have capital or permanent employment in order to obtain a bank loan. Thus, the *right* to buy is coupled with the *ability* to buy, and households lacking that ability are continuously filtered into rental flats (Andersson and Bråmå, 2004; Andersson and Magnusson Turner, 2014). In both treated areas, dwelling prices were low compared to average prices in the southern suburbs of Stockholm. These relatively cheap dwellings can serve as a first step for getting into the housing market, thus helping less well-off groups (Annadotter *et al.*, 2014). Preliminary results of ongoing research suggest that value for money was a strong motive for residents who had bought a dwelling and moved into the neighbourhoods. Respondents perceived neighbourhood quality as good enough, although they admitted there were social problems and other inconveniences.

An aspect of social equity is that authorities treat citizens equally. Both treated areas belong to the same subdivision of the Stockholm municipality administration, so investment in public spaces, public transport and other municipal services should have been equally distributed. This was not entirely the case. In treated neighbourhood A more municipal resources were allocated for upgrading green areas, playgrounds and public meeting places, such as a library. Meanwhile, in treated neighbourhood B, the municipality did not upgrade or even fully maintain the public outdoor environment over the studied period. Residents in neighbourhood A proved more capable of advocating for their needs regarding public space. Individuals or groups filed a number of citizens' propositions to the municipality and these were mainly implemented. In neighbourhood B, residents did not file any such propositions and, consequently, did not receive any allocations above the standard budget (Annadotter and Werner, 2014).

Democratic governance, interaction and trust between people are important factors for sustainable communities. According to Putnam (1993) a society's social capital and its qualifications for being a functioning democracy can be measured in terms of its number of horizontal civic associations. Although a simplified measure, associational participation is an indicator of people's will and ability to work for a common good. Neighbourhood A had higher numbers of non-profit citizen organisations than neighbourhood B. In this respect, neighbourhood A can be considered more socially sustainable than B, but this may be due to its context and history rather than the tenure transformation. The population of neighbourhood A had a higher level of education from the outset, the blocks envelop a common courtyard, each has a meeting room and other collective facilities, and the neighbourhood is better connected to the urban fabric because it is located on a subway line. All of these factors probably enhance social interaction on different levels, but they have no direct connection to tenure.

Crime is a serious problem for social sustainability, counteracting trust between people. Both treated neighbourhoods, A and B, feature in an infamous group of the fifty-five most crime-ridden and insecure neighbourhoods in Sweden

(Rikskriminalpolisen, Underrättelsesektionen, 2014). Neighbourhood B stands out with respect to arson, with numerous burnt-out cars and two nursery schools razed to the ground since 2005. In neighbourhood A, there are no such visible signs of criminal activities, but drug-dealing has been documented. In both neighbourhoods, residents have started to counteract criminal behaviour by organising neighbourhood watch groups and trying to deter young people from participating in gang culture. The groups work with the local police and the city's social services, thus linking their initiatives to the authorities.

If conversion of tenure in the treated cases had enhanced social sustainability, a development towards social equity and more sustainable communities would be expected. The case studies show some evidence of this. More data on the untreated cases will add more insight. Regarding the social equity aspect, the treated neighbourhoods differ in terms of access to daily life services and public transport, but these differences are mainly due to conditions prevailing before the conversion of tenure. In neighbourhood A, social networks for joint action have developed and have secured more municipal investment in public places and services, but the same phenomenon has not occurred in neighbourhood B. Judging from joint activities, viable social networks have started developing in neighbourhood A. In both of the treated neighbourhoods, social mix is achieved as residents with higher incomes are moving in while those of more limited means are moving out. In addition, residents' levels of education are rising, faster in A than in B. The conversion of tenure has induced change towards more resourceful residents, which in turn may enhance social sustainability at the neighbourhood level. At the regional level, increasing ownership fuels segregation as the poorest households are left with fewer alternatives.

Conclusions

The rising demand for housing in Sweden is linked to population growth and the concentration of people in the three largest cities. An inadequate supply of new dwellings, especially in the biggest cities, is, among other factors, related to scarcity and price of land for construction. The long planning process and highly demanding building regulations affect the affordability of new dwellings. Swedish buildings and infrastructure are mainly of high technical quality, designed to counteract climate change, but the immediate demand of a higher supply of affordable dwellings is yet to be met. In particular, newcomers into housing and job markets, such as young people and immigrants, face significant difficulties when trying to find a place to live. A growing inequality in housing quality is revealed in overcrowding rates, where low-income households suffer most. In the long run this development can lead to slum-like conditions and permanent social segregation, which both have serious implications for social sustainability. Another aspect of the affordability problem is that house prices have risen to a level where there is a possible price 'bubble', ready to burst and severely disturb national as well as individual financial stability. Many Swedish households carry a high burden of debt in the form of housing loans. Both social and economic sustainability are thus threatened by the lack of new construction and the lack of residential mobility.

Strict building regulations have had positive effects as the amount of energy used for heating the built environment has decreased slowly but steadily, and they are therefore successful in achieving their aims. In addition, an increasing proportion of energy is produced from renewable sources, such as water power and wind-generated power. Policies targeting households' and housing companies' energy-conserving behaviour could still help to decrease the climatic impact of housing. This is suggested by the fact that households which are able to control their indoor climates and benefit economically from saving do achieve tangible results.

The transport sector in Sweden is becoming a target for policies to reduce CO2 and harmful particle emissions – for example, taxes on fuel to decrease individual car travel and road tolls in areas with heavy traffic. The sparse Swedish population – which is scattered over many small cities and towns outside of the three metropolitan areas – poses a specific challenge for Swedish transport policy. Many of the proposals to counteract climate change are based upon urban densification, especially minimising transport and a higher level of modal split in public transport. Efficient policies will need to address less dense areas, where current public transportation is not considered a viable solution on economic grounds.

Sweden still has a strong welfare state, but housing shortages, soaring house prices and Swedish households' growing debt burden are all problematic. Policy initiatives for the housing sector are lacking. A growing public interest in housing issues is emerging and there are calls for a new political agenda for housing. The most common proposals concern house prices; lowering tax deductions on mortgage interest rates; shorter payback terms on housing loans; heavier property taxes; and higher construction rates at lower prices. The latter seems the most difficult to obtain as developers will not build without the possibility of making a reasonable profit. Building low-cost housing in the Swedish context of strict regulations and expensive land will seldom yield enough profit, so construction rates remain low. In the 1960s, strong state intervention resulted in higher construction rates and greater housing affordability, but this is not politically practicable today. At present, the political debate revolves around shortening the planning process, reforming the tax system to enhance residential mobility and modifying rental market conditions to encourage higher construction rates.

Sweden performs at a relatively high level regarding the three aspects of economic, social and environmental sustainability, although challenges remain. Economic and population growth coupled with the comparatively low risk of poverty and social exclusion are examples of Swedish society's resilience. Housing quality is generally high and the housing cost overburden rate is not particularly problematic when compared with other studied countries. Sweden also scores relatively well in terms of environmental sustainability (YCELP and CIESIN, 2014). Separately, the three aspects seem to develop in a positive direction. However, they are highly interdependent and sometimes in conflict. The issues discussed in this chapter present some examples of such conflicts.

References

English names for Swedish authorities and organisations are from official English-version websites. All title translations are by the author. All websites last accessed 7 July 2016.

Andersson, R. and Bråmå, Å. (2004) 'Selective migration in Swedish distressed neighbourhoods: can area-based urban policies counteract segregation processes?', *Housing Studies*, 19(4): 517–539.

Andersson, R., Musterd, S., Galster, G. and Kauppinen, T. M. (2007) 'What mix matters? Exploring the relationships between individuals' incomes and different measures of their neighbourhood context', *Housing Studies*, 22(5): 637–660.

Andersson, R., Bråmå, Å. and Holmqvist, E. (2010) 'Counteracting segregation: Swedish policies and experiences', *Housing Studies*, 25(2): 237–256.

Andersson, R. and Magnusson Turner, L. (2014) 'Segregation, gentrification, and residualisation: from public housing to market-driven housing allocation in inner city Stockholm', *International Journal of Housing Policy*, 14(1): 3–29.

Annadotter, K. and Werner, I.-B. (2014) *Ekonomiska investeringar i Dalen och Östberga* [Economic investment in Dalen and Östberga]. Stockholm: KTH. Available at http://kth.diva-portal.org/smash/get/diva2:773824/FULLTEXT01.pdf.

Annadotter, K., Werner, I.-B. and Gunnelin, R. (2014) *Bostadsrättsprisernas nivå och utveckling i Dalen och Östberga, åren 2005–2012* [Dwelling Prices in Dalen and Östberga – Levels and Trends between 2005 and 2012]. Stockholm: KTH. Available at http://kth.diva-portal.org/smash/get/diva2:772272/FULLTEXT01.pdf.

Boverket [National Board of Housing, Building and Planning] (2006). *Regelsamling för byggande, Boverkets byggregler* [Boverket's Regulations for Construction]. Karlskrona: Boverket. Available at www.boverket.se/contentassets/b7aa3b9e97a0408a8f0f7bbeee3c27e5/regelsamling-for-byggande-bbr-2006.pdf.

Boverket (2009) *Boendekostnader och boendeutgifter – Sverige och Europa* [Housing Costs and Housing Expenditures – Sweden and Europe]. Karlskrona: Boverket.

Boverket (2011). *Boverkets byggregler – föreskrifter och allmänna råd* [Boverket's Regulations for Construction – Binding Rules and General Advice]. Karlskrona: Boverket. Available at www.boverket.se/contentassets/a9a584aa0e564c8998d079d752f6b76d/bbr-bfs-2011-6-tom-bfs-2015-3-konsoliderad.pdf.

Boverket (2014) *De allmännyttiga bostadsföretagens utveckling och roll på bostads-marknaden 2012–2013 Regeringsuppdrag* [Public Rental Companies' Development and Their Role on the Housing Market 2012–2013 Government Commission]. Karlskrona: Boverket.

Boverket (2015a) *Behov av bostadsbyggande – Teori och metod samt en analys av behovet av bostäder till 2025* [Calculating the Need for Housing – Theory and Method Plus an Analysis of the Need for Housing up to 2025]. Karlskrona: Boverket.

Boverket (2015b) *Regelsamling för byggande* [Binding Regulations for Construction]. Karlskrona: Boverket. Available at www.boverket.se/globalassets/publikationer/dokument/2015/regelsamling-for-byggande-bbr-2015.pdf.

Boverket (2015c) *Boendesituationen för nyanlända, Slutrapport Regeringsuppdrag* [Recent Immigrants' Housing Situation, Final Report, Government Commission]. Karlskrona: Boverket.

Christophers, B. (2013) 'A monstrous hybrid: the political economy of housing in early twenty-first century Sweden', *New Political Economy*, 18(6): 885–911.

Dempsey, N., Bramley, G., Power, S. and Brown, C. (2011) 'The social dimension of sustainable development: defining urban social sustainability', *Sustainable Development*, 19: 289–300.

Energimyndigheten [Swedish Energy Agency] (2014a) *Energistatistik för småhus 2013* [Energy Statistics for One- and Two-Dwelling Buildings in 2013]. Eskilstuna: Statens Energimyndighet.

Energimyndigheten (2014b) *Energistatistik för flerbostadshus* [Energy Statistics for Multi-dwelling Buildings in 2013]. Eskilstuna: Statens Energimyndighet.

Energimyndigheten (2015) *Energy in Sweden Fact and Figures 2015*. Eskilstuna: Energimyndigheten. Available at www.energimyndigheten.se/statistik/energilaget/.

Engvall, K., Lampa, E., Levin, P., Wickman, P. and Öfverholm, E. (2014) 'Interaction between building design, management, household and individual factors in relation to energy use for space heating in apartment buildings', *Energy and Buildings*, 81: 457–465.

Esping-Andersen, G. (1990) *The Three Worlds of Welfare Capitalism*. Cambridge: Polity Press.

Eurostat (2015a) *Total Tax Revenue by Country, 1995–2014 (percent of GDP)*. Luxembourg: Eurostat. Available at http://ec.europa.eu/eurostat/statistics-explained/index.php/File: Total_tax_revenue_by_country,_1995-2014_(percent_of_GDP).png.

Eurostat (2015b) *Crude Rates of Population Change, 2012–14 (per 1000 persons) YB15 II*. Luxembourg: Eurostat. Available at http://ec.europa.eu/eurostat/statistics-explained/ index.php/Population_and_population_change_statistics.

Försäkringskassan [Swedish Social Insurance Agency] (2015) *Social Insurance in Figures 2015*. Stockholm: Försäkringskassan.

Galster, G. (2001) 'On the nature of neighbourhood', *Urban Studies*, 38(12): 2111–2124.

Galster, G. (2007) 'Should policy makers strive for neighbourhood social mix? An analysis of the Western European evidence base', *Housing Studies*, 22(4): 523–545.

Galster, G. (2012) 'The mechanism(s) of neighbourhood effects: theory, evidence, and policy implications', in van Ham, M., Manley, D., Bailey, N., Simpson, L.and Maclennan, D. (eds) *Neighbourhood Effects Research*. Dordrecht: Springer Science.

Hedin, K., Clark, E., Lundholm, E. and Malmberg, G. (2012) 'Neoliberalization of housing in Sweden: gentrification, filtering, and social polarization', *Annals of the Association of American Geographers*, 102(2): 443–463.

Hedman, L. and Galster, G. (2013) 'Neighbourhood income sorting and the effects of neighbourhood income mix on income: a holistic empirical exploration', *Urban Studies*, 50(1): 107–127.

Holmqvist, E. and Magnusson Turner, L. (2014) 'Swedish welfare state and housing markets: under economic and political pressure', *Journal of Housing and the Built Environment*, 29: 237–254.

International Monetary Fund (2011) *IMF Country Report No. 11/171, Sweden*. Washington, DC: IMF. Available at www.imf.org/external/pubs/ft/scr/2011/cr11171.pdf.

Johansson, M. and Haas, J. (forthcoming). *TOWN – Small and Medium Sized Towns in Their Functional Territorial Context*. Stockholm, Leuven: ESPON.

Kemeny, J. (1995) *From Public Housing to Social Renting: Rental Policy Strategy in Comparative Perspective*. London: Routledge.

Kemeny, J., Kersloot, J. and Thalmann, P. (2005) 'Non-profit housing – influencing, leading and dominating the unitary rental market: three case studies', *Housing Studies*, 20(6): 855–872.

Lindén, A.-L., Carlsson-Kanyama, A. and Eriksson, B. (2006) 'Efficient and inefficient aspects of residential energy behaviour: what are the policy instruments for change?', *Energy Policy*, 34:1918–1927.

Nairn, G., Gustavsson, L. and Mahapatra, K. (2010) 'Factors influencing energy efficiency investments in existing Swedish residential buildings', *Energy Policy*, 38: 2956–2963.

Putnam, R. (1993) *Making Democracy Work: Civic Traditions in Modern Italy*. Princeton, NJ: Princeton University Press.

Regeringen [Swedish Government] (2003) *En svensk strategi för hållbar utveckling – ekonomisk, social och miljömässig* [A Swedish Strategy for Sustainable Development – Economic, Social and Environmental Aspects]. Stockholm: Riksdagens tryckeriexpedition.

Regeringen (2006) *Strategiska utmaningar – En vidareutveckling av svensk strategi för hållbar utveckling* [Strategic Challenges – a Further Development of the Swedish Strategy for Sustainable Development]. Stockholm: Regeringskansliet. Available at www.regeringen.se/contentassets/b6f76a3feb8b4bb78322094dc1cdf2ba/strategiska-utmaningar---en-vidareutveckling-av-svensk-strategi-for-hallbar-utveckling-skr.-200506126.

Regeringen (2015) *Regeringens strategi inför COP21* [The Government's Strategy for Participation in COP21]. Stockholm: Regeringskansliet. Available at www.regeringen.se/artiklar/2015/04/regeringens-strategi-infor-cop21/.

Rikskriminalpolisen, Underrättelsesektionen [Swedish Police, Department of National Operations] (2014) *En nationell översikt av kriminella nätverk med stor påverkan i lokalsamhället – Sekretessprövad version* [A National Overview of Criminal Networks with Serious Effects upon Local Communities]. Stockholm: Polismyndigheten. Available at https://polisen.se/Aktuellt/Rapporter-och-publikationer/Rapporter/Publicerat---Nationellt/Ovriga-rapporterutredningar/Kriminella-natverk-med-stor-paverkan-i-lokalsamhallet/.

Söderholm, K. (2013) 'Housing, public policy and the environment in a historical perspective: lessons from Swedish post-war society', *International Journal of Sustainable Society*, 5(1): 24–42.

Statistiska Centralbyrån (SCB) [Statistics Sweden] (2015a) *GDP: Production Approach (ESA2010), by Industrial Classification SNI 2007*. Stockholm: SCB. Available at www.statistikdatabasen.scb.se/pxweb/en/ssd/START__NR__NR0103__NR0103E/NR0103ENS2010T06A/?rxid=c4e325f3-7151-42b3-9078-ee60457efb71.

SCB (2015b) *Population and Population Changes in Sweden. Year 1749–2014*. Stockholm: SCB. Available at www.statistikdatabasen.scb.se/pxweb/en/ssd/START__BE__BE0101__BE0101G/BefUtv1749/?rxid=c4e325f3-7151-42b3-9078-ee60457efb71.

SCB (2015c) *Population Changes Number of Persons by Region and Sex. Year 2000–2014*. Stockholm: SCB. Available at www.statistikdatabasen.scb.se/pxweb/en/ssd/START__BE__BE0101__BE0101G/Befforandr/?rxid=c4e325f3-7151-42b3-9078-ee60457efb71.

SCB (2015d) *Population Density per sq. km, Population and Land Area by Region and Sex. Year 1991–2014*. Stockholm: SCB. Available at www.statistikdatabasen.scb.se/pxweb/en/ssd/START__BE__BE0101__BE0101C/BefArealTathetKon/?rxid=f49e6b07-cc93-4820-95a9-209b2457f781.

SCB (2015e) *Population by Age and Sex. Year 1860–2014*. Stockholm: SCB. Available at www.statistikdatabasen.scb.se/pxweb/en/ssd/START__BE__BE0101__BE0101A/BefolkningR1860/?rxid=7ddd5a4e-90a7-42d8-9e7d-13a36da08d0a.

SCB (2015f) *Number of Dwellings by Region, Type of Building and Period of Construction. Year 2013–2014*. Stockholm: SCB. Available at www.statistikdatabasen.scb.se/pxweb/

en/ssd/START__BO__BO0104/BO0104T02/?rxid=7ddd5a4e-90a7-42d8-9e7d-13a
36da08d0a.

SCB (2015g) *Converted Dwellings in Multi-dwelling Buildings by Region, Type of Ownership, Period of Construction and Size of Dwelling. Year 1989–2014*. Stockholm: SCB. Available at www.statistikdatabasen.scb.se/pxweb/en/ssd/START__BO__ BO0102__BO0102B/LagenhetOmbAkBpLtAr/?rxid=7ddd5a4e-90a7-42d8-9e7d-13a36da08d0a.

SCB (2015h) *Dwellings in Newly Constructed Buildings by Region and Type of Building. Quarterly 1975K1–2015K3*. Stockholm: SCB. Available at www.statistikdatabasen.scb. se/pxweb/en/ssd/START__BO__BO0101__BO0101A/LagenhetNyKv/?rxid=ec412e5b-8210-4e79-9d03-f72d782e9bc7.

SCB (2015i) *Housing by Indicator, Age and Sex. Percentage and Estimated Numbers in Thousands. Year 2008–2009 –2014–2014*. Stockholm: SCB. Available at www. statistikdatabasen.scb.se/pxweb/en/ssd/START__LE__LE0101__LE0101B/LE0101 B01/?rxid=6fac9ee9-7624-4a3a-acd8-c987c5ce8d7e.

SCB (2015 j) *Genomsnittlig boendeutgift, bostadsbidrag och boendeutgiftsprocent per hushåll (HEK) fördelad efter region, upplåtelseform och hushållstyp. Urvalsundersökning, se fotnoter. År 2004–2013* [Average Housing Expenditure, Housing Benefits and Proportion of Housing Cost/Income per Household, per Region, Tenure and Type of Household. Sample Survey. Year 2004–2013]. Stockholm: SCB. Available at www.statistikdatabasen. scb.se/pxweb/en/ssd/START__HE__HE0103__HE0103E/Boendeutgift/?rxid=836db7c1-36da-4176-90bd-e3c50980e5e1.

SCB (2016) *Number of Dwellings by Region, Type of Building and Tenure (Including Special Housing). Year 2013–2015*. Stockholm: SCB. Available at www.statistikdatabasen.scb.se/ pxweb/en/ssd/START__BO__BO0104/BO0104T04/?rxid=bea338eb-6c21-44e0-94d4-3b72871cf0b1.

Sveriges Byggindustrier [Swedish Construction Federation] (2014) *Fakta om byggandet 2013* [Facts on the Annual Construction 2013]. Stockholm: Sveriges Byggindustrier. Available at https://www.sverigesbyggindustrier.se/.

Sveriges Riksbank [Central Bank of Sweden] (2014). *Hur skuldsatta är de svenska hushållen?* [How Much Debt Do Swedish Households Carry?]. Stockholm: Riksbanken. Available at www.riksbank.se/Documents/Rapporter/Ekonomiska_kommentarer/2014/ rap_ek_kom_nr01_140507_sve.pdf.

Sveriges Riksbank (2015) *Finansiell stabilitet* [Financial Stability]. Stockholm: Riksbanken.

Werner, I.-B. and Annadotter, K. (2013) *Demografiska förändringar i Östberga, Dalen, Bredäng, Söderort och Stockholms kommun – 1990–2011*. [Demographic Changes in Östberga, Dalen, Bredäng, Southern Suburbs and Stockholm City, 1990 to 2011]. Stockholm: KTH.

World Commission on Environment and Development (WCED) (1987) *Our Common Future*. New York: UN-WCED. Available at www.bne-portal.de/fileadmin/unesco/de/ Downloads/Hintergrundmaterial_international/Brundtlandbericht.File.pdf?linklisted= 2812.

Yale Center for Environmental Law and Policy (YCELP) and Center for International Earth Science Information Network (CIESIN) (2014) *2014 EPI, Country Rankings*. New Haven, CT: YCELP and CIESIN. Available at http://epi.yale.edu/epi/country-rankings.

5 Norway

Eli Støa

Introduction

From being a poor, mainly agrarian country with a dispersed population, Norway has had steady economic growth since the early twentieth century, due to revenue from the large merchant fleet and new energy-intensive industries based on hydro-electricity. Urbanization started relatively late compared to many other European countries, mainly in the twentieth century, and it could be claimed that, even today, Norwegians have a rather ambivalent relationship to urban life (Aune and Støa, 2016).

Since the 1970s, the country has enjoyed increased income from off-shore oil and gas fields and a GDP growth of almost 180 per cent over the last 45 years. The GDP in 2014 was nearly 80 per cent above the EU average (Statistics Norway, 2015a). Although the Global Financial Crisis (GFC) of 2008 had some minor consequences for Norwegian industry and the construction sector, its effects were limited (Vestergaard and Haagerup, 2016). Unemployment in Norway is low. From the 1970s until mid-1980s it was quite constant at 2 per cent, but increased to between 5 and 6 per cent in the 1990s. In 2015, it was 4.5 per cent (Statistics Norway, 2015b). During the first half of 2015, there was a rise in unemployment linked to redundancies in the oil sector. By December 2015, the economic situation was marked by a higher degree of uncertainty than in 2008 (Statistics Norway, 2015c). Income inequality is lower in Norway than for most other European countries, although it has increased slightly since the 1980s (Statistics Norway, 2014). The highest share of persons with a low income is found in the capital, Oslo, and the most dispersed municipalities of the country.

The Norwegian population was approximately 5.2 million in 2015, an increase of more than 1.5 million since 1950 (Statistics Norway, 2015d). 13.4 per cent of the population are immigrants (Statistics Norway, 2016). In Oslo, immigrants constitute one-quarter of the population (Statistics Norway, 2016). As in other European countries, the number of elderly people is growing and household size is continuously decreasing (from 3.4 persons per household in 1946 to 2.1 in 2014). Four out of ten Norwegian households consist of just one person (Statistics Norway, 2015a), and in the capital this rises to over one in two households (Statistics Norway, 2012).

Social democratic governments and welfare state ideals dominated the first decades after the Second World War until the 1980s. Thereafter, there was a shift towards more neoliberal policies with increased privatization, even though social democrats have been in government for most of the period. Norway applied for EU membership in both 1972 and 1994. However, the agreements were rejected in national referendums on both occasions.

Although Norway scores rather high on the Environmental Performance Index (EPI) (see Table 1.1), the country still has major challenges in order to reach its obligations when it comes to international climate agreements. The government has set as a national objective becoming a 'carbon neutral' society by 2050 (Norwegian Ministry of and the Environment, 2012). In 2013, GHG emissions were 4.6 per cent higher than in 1990 (Norwegian Environment Agency, 2014) and, while they have decreased gradually every year since 2010 (Norwegian Government, 2015), there is some way to go to achieve carbon neutral status. The transport sector is one of the main sources of emission of climate gases. Its share of emissions has increased by 30–40 per cent since 1990 (Statistics Norway, 2013a), and private cars are the main source. During the same period, the share of road traffic has been stable at around 80 per cent of all transport, while public transport (buses, trams and trains) accounts for only 11 per cent (Statistics Norway, 2013a). Energy consumption has increased by almost 50 per cent since 1976. When measured per capita, Norway is slightly above the average for European countries. The household sector represents a substantial component of total energy consumption, and while it has decreased per household and per square metre of dwelling space, stationary energy use in total increased by approximately 12 per cent from 1990 to 2009 (Bøeng *et al.*, 2011).

Most political parties agree that in order to reach the climate goals, a fundamental transition from an economy based heavily on oil and gas to one that promotes more environmentally friendly industries is needed (Norwegian Ministry of Climate and Environment, 2014). Urban planning and housing have been highlighted in policy documents, including reducing environmental impacts from the transport sector and from the operation of buildings. There is, however, less focus on how socio-cultural practices, such as residential consumption patterns and mobility habits, will affect the future urban environment. These issues, and how they relate to urban planning and building design, are discussed later in this chapter. First, the chapter outlines the main characteristics of the Norwegian housing landscape, and then presents an overview of how sustainable cities, communities and housing are dealt with in relevant policy documents.

Housing in Norway

During the fifties and sixties, large housing estates were built in and outside the major cities based on functionalist ideas. The estates catered for the growing number of people moving into the cities, and provided improved living standards for those who had been residing in old and rundown working-class housing. Much of the wooden house stock in urban quarters was demolished to make room for medium- to high-rise apartments. The private car became a common commodity during the 1960s, contributing to increased urban sprawl. For those who could afford it, fields of catalogue houses were constructed in the outskirts of the towns and cities. Subsidized mortgages from the State Housing Bank made this sprawl a popular alternative.

The modern, rationalistic planning regime met increased resistance at the end of the 1960s and 1970s. Alternative movements emerged which emphasized community and the preservation of historically valuable areas. Instead of high-rise housing estates, low and dense areas became the standard for multi-family housing. Since the 1980s, there has been a growing interest in urban lifestyles and urban housing. Centrally located industrial urban areas (often docklands) were released for housing purposes. The development of brownfield sites took place and was supported by a densification policy according to compact city strategies and sustainable urban development concepts.

In 2015, eight out of ten Norwegians lived in an urban settlement (defined as a collection of houses with at least 200 inhabitants and fewer than 50 metres distance between houses), compared with 50 per cent in 1945 (Statistics Norway, 2015a) and more than one-third of the population lives in one of the five cities that have more than 100,000 inhabitants (Statistics Norway, 2015e). There are more than 2.4 million dwellings in the country, almost 80 per cent of which are one–two-storey wooden buildings. These may be detached houses, semi-detached or terraced units. Flats comprise only 22 per cent of the housing stock.

Photo 5.1 Dalen Hageby, a low-density housing area in Trondheim built in 1961
(Architect: Jarle Øyasæter).

Photo: Eli Støa.

Since the Second World War, the housing sector in Norway has consisted of three main actors: the state, municipalities and the private sector. These actors have different roles and responsibilities and the interrelationship between them has changed over time. Initially, the responsibility of the state was to provide subsidized loans to the general public for housing construction via the Norwegian State Housing Bank. The bank was established in 1946 and played a dominant role in housing finance in Norway for most of the postwar period. In addition to setting requirements related to cost and design standards, the bank played a vital role in securing the quality of housing stock. The municipalities provided land, plans and necessary infrastructure, while the executive role was left to the private sector. The overall political goal was to provide individual ownership, and today almost eight out of ten households are homeowners (Statistics Norway, 2015a), including about 20 per cent in housing cooperatives.

A housing cooperative is an ownership model where the residents share the ownership of the buildings and the site. Cooperative housing played an important role during the first decades after the Second World War, especially in towns and cities. During this period, the term 'social housing' was understood as cooperatively owned with state bank finance and price regulation. Today, the term covers only the rather limited stock of public rental housing. In 2011, this amounted to approximately 100,000 units in total, a number which constitutes a little more than 4 per cent of the total housing stock.

The private sector's impact on planning and construction started to increase in the 1980s. The State Housing Bank now competes with private banks on the mortgage market and there are no price regulations on cooperative housing. Social housing policy is currently directed towards selected groups defined as disadvantaged in the housing market. According to the Norwegian government, these groups constitute less than 3 per cent of the population (Norwegian Ministry of Local Government and Modernization, 2014). They include immigrants, low-income families, people with disabilities or mental health problems and homeless people. In addition to the provision of public rental housing, current public social supports for disadvantaged groups consist of low-interest loans and subsidy schemes (housing allowances and grants) given on individual bases (Sørvold, 2011).

Within a current market-driven housing sector, public authorities have lost their central role and the measures that are needed to guide housing development. In the present economic situation it is argued that housing production is too low as demand is high, especially in urban areas. The result is expensive dwellings due to high demand and the high cost of land. House prices in Norway have generally had a higher overall growth rate than those in many other European countries (Vestergaard and Haagerup, 2016). After a drop in 2008, prices have increased by 5–7 per cent annually (Vestergaard and Haagerup, 2016).

Norway has some of the highest housing standards in Europe. Nearly all dwellings have an indoor toilet, bath and separate kitchen, and about one-third have two or more bathrooms (Andersen, 2004). Eight out of ten people have access to their own garden (Statistics Norway, 2002a). More than half of

Norwegian dwellings are detached houses and the average living space per person was almost 60 square metres in 2013 (Statistics Norway, 2015a). In addition, Norway has the highest frequency of second-home ownership in the Nordic countries (Müller, 2007), with almost half of Norwegian households having access to a second home (Støa and Manum, 2013). According to a Norwegian survey on living conditions, only 6 per cent of Norwegians live in overcrowded dwellings, while 8 per cent stated that they live in dwellings with problems such as moisture and/or decay (Statistics Norway, 2013b). The proportion living in overcrowded dwellings has been quite stable since 1980, while the group with more than three rooms per person has increased (Sæther, 2010). Smaller dwellings and overcrowding are both more common in the larger cities, among households on social benefits and among non-Western immigrants (Sæther, 2010). For the country as a whole, more than half of non-Western immigrants live in overcrowded dwellings, while the corresponding figure in Oslo is 66 per cent (Statistics Norway, 2002a).

Sustainable communities and housing

As noted earlier, reducing the environmental impacts from the transport sector and the operation of buildings is central to Norway's goal to become a low carbon society. Key strategies include urban densification and the development of compact cities, where people can easily reach their everyday activities without using private cars. There have also been efforts to combine these strategies with more holistic approaches to sustainability through large government initiatives.

Compact cities as the main sustainability strategy

Sustainability has been high on the agenda in Norwegian policy documents, planning strategies, research and public debates since *Our Common Future* was published in 1987 (WCED, 1987). One reason for this may be the fact that the commission was led by a former Norwegian prime minister, Gro Harlem Brundtland. Prior to its publication, the environmental debate had emphasized isolated issues such as pollution and energy demand, but this has been extended now to include the relationship between people and nature, present and future, and the distribution between poor and rich (Guttu, 2003). Critical voices within the Norwegian environmental movement raised questions about the philosophy of growth which had been a foundation of the social democratic welfare state. They questioned the notion that the main environmental challenges could be solved by technological innovation, emphasizing instead the significance of increasingly excessive consumption patterns. From this perspective, housing could no longer be regarded as only a basic right, since most Norwegian had reached the level of acceptable housing standards, but also as a threat to the environment (Guttu, 2003: 467). Postwar housing policy had led to a growth in living space and energy demand for heating and transport partly due to the large proportion of detached houses and urban sprawl. Between 1970 and 1990, the use of land per person in urban areas increased by almost 25 per cent (Næss *et al.*, 2015: 37).

After 1987 there was a shift in emphasis from a rather one-sided focus on building technology and energy-efficient building envelopes to more urban planning issues such as land use, transport and green structure. In the Nature and Environmentally Friendly Urban Development (NAMIT) project (1988–1992), eight research institutions worked together to translate the conclusions of the Brundtland Report into planning strategies and development patterns. NAMIT initiated *compact cities* as a model for sustainable urban development in Norway (Hanssen *et al.*, 2015a). The project marked a change from the postwar functionalist and car-based urban planning. It was followed by a governmental programme on Environmentally Friendly Cities (Norwegian Ministry of the Environment, 2000) where five cities were selected as 'urban laboratories' (Næss *et al.*, 2015). This program, lasting from 1993–2000, took an even more extended and integrated approach to sustainable urban development along the following lines: Land use and transport; development of urban centres; urban housing; vibrant local communities; green structure; and urban design and cultural heritage (Norwegian Ministry of the Environment, 2000). It was taken for granted that urban densification would have environmental benefits. The recommendations of the programme dealt with how to secure economic viability and quality of life within compact cities. Important issues were social and functional diversity, involvement of private and public stakeholders, developing attractive public spaces and securing housing qualities in a broad sense (Norwegian Ministry of the Environment, 2000).

During the late 1990s and early 2000s, concerns began to emerge among planners, environmental associations and researchers about possible negative implications of densification (e.g. Skjeggedal, 1996; Duun, 1996; Holden and Norland, 2005). The social dimension of the sustainability concept was brought forward, often with reference to the Danish concept of Urban Ecology (FBBB, 2010), where the local context and residents were at the centre. The concept includes the way we build as well as the way we live, our use of resources, everyday behaviour and the way we plan our societies (FBBB, 2010). Experimental urban neighbourhoods were highlighted as examples of alternative approaches to sustainable development where user involvement and local community were emphasized (e.g. By- og Boligministeriet, 2001; Svane and Wijkmark, 2002; Støa, 2009).

In 2008 a new government programme, Cities of the Future, replaced Environmentally Friendly Cities (Norwegian Ministry of Local Government and Modernization, 2015a, 2015b). Several ministries were involved and thirteen of the largest Norwegian cities took part, encompassing approximately 36 per cent of the Norwegian population (Haagensen, 2015). The overall goal of the new programme was to reduce total GHG emissions from road transport, energy use in buildings, consumption and waste in urban areas and at the same time develop strategies for addressing future climate change (Norwegian Ministry of Local Government and Modernization, 2015a, 2015b). The aim was to reduce GHG emissions in the participating cities by 35 per cent by 2030, and by 24 per cent by 2020 (Selvig and Opheim, 2009). One of the priorities was to develop an improved urban environment from social, economic and environmental perspectives. The following were highlighted: a healthy and vibrant city; a city for pedestrians, bicycles and public transport; a city with urban spaces and meeting places; and a blue–green city (Selvig and Opheim, 2009: 7). Cities of the Future should:

1   contribute to high quality of life for all inhabitants;
2   use their own character and advantages as starting points;
3   offer high-quality services and opportunities for all; and
4   make it easy to live a climate-friendly everyday life.

(Framtidens byer, 2012: 6)

In addition to indicators that relate directly to environmental topics, such as land use, energy use, emissions, local pollution, modes of transport and energy sources, the programme includes indicators that relate to social issues, such as safe access to playgrounds and recreational areas, proportion of shops in urban centres, distance to schools, day care and grocery shops (Haagensen, 2015).

Surveys of the inhabitants of the thirteen cities during the programme period assessed their attitudes towards the environmental plans of their municipalities, and their own role as contributors to an improved environment (Rambøll, 2015). The results indicated that the inhabitants experienced an improved and strengthened focus on the urban environment, particularly on issues relating to consumption and waste. It is also worth noting that the respondents claimed to have gained increased understanding of how their own behaviour might influence climate change, and that they perceived fewer obstacles in the local community to living a more environmentally friendly life (Rambøll, 2015: 32). However, reducing GHG emissions proved somewhat more problematic. While emissions per inhabitant decreased by 2.5 per cent during the period 2009–2012, they increased by 1.5 per cent in the Cities of the Future counties over the same period (Haagensen, 2015: 70). These results indicate that the goal of 24 per cent reduction by 2020 may be very optimistic. Nevertheless, both the Cities of the Future and the Environmentally Friendly Cities programmes represent a shift away from the functionalist, car-based planning regime that characterized much of the second half of twentieth century and towards more sustainable cities.

Sustainable communities – the social turn

While definitions of 'sustainable communities' vary considerably in the literature (Winston, 2013), the concept generally encompasses two main elements: first, the need for integrated solutions combining all three pillars of the sustainability concept; and, second, within this holistic approach, an emphasis on the social pillar because of the inclusion of the term 'community'. Social sustainability implies securing a society's social and cultural viability over the long term on both collective and individual levels. There have been several attempts to define 'social sustainability' through identification of key themes or indicators (e.g. Murphy, 2012). In the context of urban form, Bramley *et al.* (2009: 2126) suggest that social sustainability comprises two main dimensions: social equity issues (access to services, facilities and opportunities); and sustainability of community (place attachment, social interaction, safety/security, perceived quality of local environment, satisfaction with the home, stability, participation). While these may be regarded as societal goals in their own right, the focus in this chapter is on how social sustainability may contribute to achieving environmental goals. As argued by Vallance *et al.* (2011: 344), 'it is a little unrealistic to expect people to care about global warming or species extinction when they are cold, hungry, seeking work, or feel unsafe in their own home'. Even if Norway has increasing challenges in terms of social inequality and housing quality, the main challenge is unsustainable consumption patterns and a housing culture where individual freedom and control are highly valued, leading to resource-consuming residential patterns with a high share of car-based detached houses (Støa and Aune, 2012).

While 'sustainability' is widely used in relation to housing, urban development and the construction sector, the term 'sustainable communities' is not much present in Norwegian policy documents. However, it is very much embedded in one of the instruments established by the Ministry of Local Government and Modernization (KMD) in 2014 as part of its latest urban and community development initiative (Norwegian Ministry of Local Government and Modernization, 2015c). The instrument, the Urban Neighbourhood Renewal programme, aims

Photo 5.2 One of the selected neighbourhoods for the governmental Urban Neighbourhood Renewal programme is Saupstad-Kolstad in Trondheim.

Photo: Ole Kristian Lundereng © Trondheim Municipality.

to improve the environment, housing and living conditions in three selected areas through cooperation between the state and the municipalities. At the time of writing, three neighbourhoods had been selected: Groruddalen (Oslo), Saupstad-Kolstad (Trondheim) and Årstadvollen (Bergen). All three were constructed in the 1960s and 1970s and they all face social challenges, including high proportions of residents on social benefits and non-Western immigrants.

In presenting the programme, the Norwegian State Housing Bank defined sustainable development by emphasizing the physical and social factors regarded as requirements for a well-functioning local community:

> A sustainable development is a robust and long-term process, in which physical and social factors are viewed in context within a given area. A good living environment should, amongst other things, include physical conditions such as universally accessible meeting venues, play areas, good housing and green areas, as well as maintain residents' sense of safety and belonging. A varied and easily accessible range of services based on the residents' needs is also important for a well-functioning local community.
>
> (Norwegian Ministry of Local Government
> and Modernization, 2014: 28)

The programme identifies six areas on which performance will be assessed. These areas give the most comprehensive overview of what are regarded as indicators for sustainable communities in Norway (see Box 5.1).

Box 5.1   Indicators of sustainable communities in Norway

I. Housing and building quality

- Improved accessibility and energy standards
- Increased use of governmental instruments to increase quality
- Increased housing diversity
- Upgrading of technical and aesthetic standards of buildings

II. Residential environment

- Upgraded and new meeting areas for children and adults
- Reduced negative environmental impacts
- Outdoor areas made accessible for all, technically and aesthetically upgraded
- Increased perception of safety in outdoor areas as well as on walking and cycling paths
- Increased feelings of ownership and belonging to the neighbourhood

III. Children and young people

- Accessible, upgraded and new social arenas for children and young people
- Children and young people should perceive their neighbourhood as safe
- Establish preventive cooperation between municipality, school/ kindergarten and voluntary associations

IV. Participation and involvement

- Increased number of do-it-yourself activities in the neighbourhood
- Increased participation in cooperative residential activities
- Increased number of members of and increased activity level in local organizations
- Establish a coordination committee for different local organizations
- Establish new voluntary associations and organizations
- Mobilize local youth as leaders and role models
- Residents' participation in planning processes

V. Local and accessible services

- Establish new public services and outreach programmes
- Establish new private services and local shops
- Improve existing public and private services
- Improve coordination of public services

VI. Perceived housing quality

- Documentation of increased perception of safety and well-being in the neighbourhood
- Increase number of residents living in suitable dwellings
- Increase number of residents who state that they intend to remain in the neighbourhood

(Norwegian State Housing Bank, 2014: 8–9)

A review of policy documents dealing with sustainability and housing shows that Norwegian authorities are acutely aware of the need for a holistic approach to sustainable development, including all three pillars. This is particularly the case with the Urban Neighbourhood Renewal programme. However, the latter is a rather limited programme that targets only three neighbourhoods, and it seems that when it comes to general planning and construction practice, there is still a reluctance to take into account qualitative issues that may affect community life over the long term.

Barriers, drivers and challenges for sustainable communities in Norway

Research on the impact of the compact city strategy to date has concluded that it led to a shift in urban development from increased land use per person until the 1980s to reduced land use in the early 2000s. Transport by private car is still increasing, but Næss *et al.* (2015) argue that the growth is less than it would have been without the densification policy, and transport statistics show that the share of travel to work by car in the Oslo region declined from 59 per cent in 2007 to 46 per cent in 2012.

Developments regarding energy demand in the building sector are somewhat ambiguous. The emphasis is on technical improvements and innovation as well as stricter planning and building requirements. This has led to a reduction in energy demand per square metre, but stationary energy demand in the household sector has continued to increase due to, for example, more living space per person and increasing comfort levels (Bøeng *et al.*, 2011).

Even if it has proven hard to fulfil all ambitions within the environmental pillar of sustainability, it may be argued that Norway is slowly moving in the right direction. It seems, however, that 'various social considerations' are the losers in compact city development (Hanssen *et al.*, 2015b: 262). These considerations relate to both social equity and sustainability of community, including housing quality (Hofstad, 2015; Schmidt, 2015). Even though the Norwegian housing sector is characterized by high standards when it comes to quantitative aspects, such as private space and technical characteristics, there has been increasing criticism of the architectural quality of private and public spaces in urban residential areas (e.g. Støa *et al.*, 2006; Guttu, 2008; Guttu and Schmidt, 2012; Schmidt, 2015). Apartments are becoming smaller and some lack fundamental qualities, such as sufficient daylight, views, privacy and adequate space for storage. Green and recreational spaces have become increasingly fragmented and diminished. Private residences turn their backs on city life instead of contributing to its activities, resulting in a neglect of community and public spaces. Building layout and construction lack the necessary flexibility to be sufficiently robust to adapt to various and changing households and lifestyles over time (Støa, 2012). In addition, there has been increasing social segregation in some urban areas due to an unbalanced composition of dwelling size, type and price levels (Støa *et al.*, 2006; Schmidt 2015). A high demand for urban housing has led to a significant increase in house prices and rental costs over recent decades (Vestergaard and Haagerup, 2016), in spite of the low standards. Low-income groups have had to move out of the city centre to suburban housing estates where housing costs are lower as the urban apartments in general are too expensive for them.

Barriers

The main barriers to sustainable communities in Norway are socio-cultural, although these are clearly intertwined with both economic and political barriers. As mentioned above, Norway does not have a long urban history. Most Norwegian cities are rather small, and the suburban single-family house still seems to be regarded as the ideal home by most Norwegians. In spite of public campaigns and programmes that aim to strengthen urban qualities, it seems that city life cannot compete with the individual freedom, privacy and closeness to nature that detached houses can offer (Støa, 1996). This resource-demanding housing culture is not only a question of values but also about a general economic affluence. With strong economic growth during the twentieth century, Norwegian consumption patterns are characterized by large private space, small household size, second homes, high rates of car ownership, and a preference for individual solutions. Dwellings are increasingly regarded as investments, and less as life projects.

Thus, it seems that in spite of the potential opportunities that a rich country like Norway has to create a durable, high-quality built environment, the construction sector appears reluctant to test sustainable solutions combining social and environmental innovations. This includes pioneering technical solutions that may require high investment costs and where demand is not yet established. It may be particularly true for socially innovative housing solutions, where the emphasis is on community and the relationship to the city rather than on private space only, with a high degree of flexibility, collective and/or intergenerational solutions. Here, lack of innovation may be due to insufficient demand as Norwegian home-buyers tend to be rather conservative. In addition, there is a lack of governmental support and research funding for projects that focus on qualities other than technical ones.

With the increased liberalization of the housing market there is a lack of policy instruments in this area. Improving housing for disadvantaged groups – such as young people, the homeless, the mentally ill, low-income groups and other marginal groups – is dealt with mainly by individual social benefits rather than as part of an active urban housing policy. Qualities of the built environment, secured by legal requirements, are more or less limited to what may be measured, such as energy demand and accessibility for people with disabilities, while other qualities relating to, for example, aesthetics, usability and community are left to private actors.

Current housing policies are characterized by socio-economic interventions targeted at disadvantaged groups, on the one hand, and technical and functional interventions, on the other. Few attempts have been made to combine the two in an integrated way. The Cities of the Future programme, which took a more holistic approach to sustainable development issues, and the more recent Urban Neighbourhood Renewal programme are exceptions. However, it is still too early to assess these programmes' impact on the general urban planning and housing market.

Drivers

In spite of the barriers mentioned above, there are clearly some positive trends, too. I have already mentioned that the compact city strategy has resulted in urban areas moving slowly in a positive direction towards reduced emissions from transport and less land use per person. Important drivers of this have been a combination of economic profitability of development of inner-city lots such as vacant brownfield sites, particularly in dockland areas; stronger preference for more urban lifestyles; political will; and overall agreement among planning actors that densification is the best option (Hanssen *et al.*, 2015a, 2015b; Norwegian Ministry of Local Government and Modernization, 2015c).

Norwegian politicians generally acknowledge that solving climate and environmental challenges is one of the greatest responsibilities of our time (Norwegian Ministry of Climate and Environment, 2014). A majority in parliament agreed on a climate settlement in 2008 and 2012 which stated that Norway should be carbon neutral by 2050 (Norwegian Ministry of the Environment, 2008, 2012). This remains valid, despite the conservative Progress Party (the only political party that refused to sign up to the settlement) joining the government in 2013.

In addition to political support for densification and more environmentally friendly transport solutions, the politicians' focus on climate challenge has led to a stronger push, and more funding for research and technological developments, particularly with respect to renewable energy and energy efficiency. Several centres for environmently friendly energy research have been established, with funding from the Norwegian Research Council and industry partners. One of these is working on zero emission buildings (ZEB, 2015). Along with several other initiatives, these have led to increased interest from the building sector. It seems to be generally acknowledged among the leading actors that this needs to be taken seriously and that it may provide profitable new business opportunities (e.g. FutureBuilt, 2015; Green Building Alliance, 2015).

Public attitudes about the need to act on environmental challenges are slowly improving. The 'Climate Barometer' is an annual survey that asks people for their views on environmental issues. In 2015 respondents rated 'climate change' second on the list of Norway's most important societal challenges, up from sixth five years earlier (TNS Gallup, 2015). The Green Party was elected to the Norwegian parliament for the first time in 2013, although with only one representative. In local government elections in September 2015 it achieved a 'breakthrough', with almost 5 per cent of the votes, which resulted in an increase in its elected representatives from 30 to 285. Despite these trends, however, the Climate Barometer showed that, while many Norwegians appreciate the seriousness of the matter, few are willing to make changes to their daily lives and habits (TNS Gallup, 2015).

Challenges

It is indeed a challenging task to combine current levels of income with national goals to reduce climate gases and meet international obligations. A strong and continuously growing economy has led to unsustainable consumption patterns which are hard to alter in an affluent society. The political economy is based on growth and there is generally a lack of political will to question this from a sustainability perspective.

After a period when the social pillar was foregrounded, for example in various government programmes, the focus now appears to be back on the environmental dimension of sustainability. There is still a tendency to see technological development and compact city strategies as separate from social sustainability, despite good intentions. Social and qualitative aspects seem to get lost in the process from idea to realization of urban projects and there are limited examples of how to translate the vision of a carbon-neutral society into meaningful action at the local level, with very few projects managing this transition (Støa *et al.*, 2014).

A major challenge is to identify potential conflicts between our current way of life and what may prove necessary in order to reduce the carbon footprint of modern societies to within safe levels (Støa, 2014). We need to create conditions under which shifts in values can take place without being restrained by material surroundings that afford a limited range of possibilities (Larssæther *et al.*, 2014). Urban planning and architecture may play important roles in addressing this challenge, as is elaborated in the next section.

Towards more sustainable residential communities – lessons from the Brøset project

The construction of sustainable communities requires residential cultures marked by sensitivity towards the environment. 'Residential culture' is here understood as a social practice involving not only individual households' everyday choices, but also the dynamics between the built environment, households' skills and competences and images of and meanings relating to ideal homes and residential areas (Støa and Aune, 2012). Practices are reconstructed and transformed continuously in an interactive process among the various actors (Shove *et al.*, 2007). Looking back at the recent history of housing proves that residential cultures have gone through remarkable changes over the last century, in terms of household structure, everyday life in dwellings and the technical installations in houses (Grytli and Støa, 1998). What is often taken as a given – 'the Norwegian ideal home' and the wish for ever-increasing space and belongings – is perhaps not as stable as it might seem. This opens up the possibility for change and for more sustainable residential practices than the ones that are often taken for granted (Støa and Larssæther, 2014). Røe (2015) argues that architects play a central role when the compact city is to be visualized and concretized in specific settings. Using visual representations of how everyday activities and social life may turn out, the idea of densification may be justified and appear attractive. This can be seen as a kind of marketing of an idea, on one hand, but also as a way of envisioning an unknown future and thus as a means to initiate debates and new visions of urban life. In this section I will briefly look at the roles that planners and architects may play in this context, using the planning process for a new urban neighbourhood in Trondheim – the Brøset project – as an example (Støa *et al.*, 2014).

The Brøset case

In autumn 2007, the municipality of Trondheim selected the thirty-five-hectare Brøset area for development as a sustainable urban neighbourhood. 'Sustainable' included low energy demand and healthy materials as well as social and economic issues, such as low-cost housing for disadvantaged groups. The ambition was to move beyond the conventional focus on technology, materials and energy use, and involve issues connected to lifestyle, such as daily transportation, leisure-related travel and consumption of goods and services. The vision was to create a neighbourhood in which residents could live, work, shop, go to school and find meaningful leisure activities without generating an unsustainable carbon footprint (Trondheim City Planning Office, 2010).

The project played a central role in the city of Trondheim's contribution to the previously mentioned national policy initiative Cities of the Future (Norwegian Ministry of Local Government and Modernization, 2015a, 2015b). A planning programme for the Brøset area was approved by the City Council in April 2010, where the goal was established to reduce the carbon footprint per person over the long term from 8–11 tons to 3 tons CO_2 equivalents per year (Trondheim City Planning Office, 2010; Solli and Bohne, 2014). During summer 2010, four interdisciplinary teams were selected to draft proposals for the design and further development of the area. The outcome of this process was presented to the public in January 2011 and used as a basis for the development of the zoning plan for the Brøset area. This was finally approved by the city authorities in summer 2013.

At the time, the Brøset project was among the most ambitious urban development plans in Norway. It combined relatively high housing density (13 dwellings/ha) and functional mix with a large proportion of green space for recreation, sports and urban farming. The area was divided into eleven neighbourhoods, each consisting of 60–300 housing units and a community building. Each neighbourhood was to be diverse in terms of dwelling sizes in order to provide a mix of households. Apart from this, no other measures were taken to facilitate access for disadvantaged groups. Walking and cycling were prioritized as the main transport modes. The number of parking lots was dramatically reduced compared to other residential neighbourhoods in the same area, and public transport planned to be easily accessible. Stationary energy was to be provided from local renewable sources. Furthermore, extended participation among potential residents was to be implemented in the future. The zoning plan and the intentions presented in the planning programme would, according to the planning regulations, be expanded into a more specific environmental programme (Trondheim City Planning Office, 2013).

At the time of writing, the construction work in the Brøset neighbourhood had not yet started. It is still uncertain whether the site owners have reached an agreement and will pursue the plans, thus demonstrating that the city has not undertaken an easy task. One of the reasons for this is that during the planning process (which took almost eight years), scepticism among the main local actors in the construction sector increased. After initial enthusiasm about participating in

Photo 5.3 Aerial view of the Brøset area from the south.

Source: M. Herzog /www.visualis-images.com.

a future-oriented development, they started to question if there was a market for this kind of housing (e.g. Hansen, 2012). Nevertheless, there are important lessons to be learned from the process so far, including the role of architecture and planning. I briefly present three of these below (adapted from Larssæther *et al.*, 2014).

Creating an image of new normalities

A major aim set by the city of Trondheim was that the four selected teams should provide design solutions that would make it easy for residents of Brøset to live 'low-emission lives' (Trondheim City Planning Office, 2010). This resulted in proposals that in many ways differed substantially from what is currently available in the Norwegian housing market. Elements that were particularly noticed by the professional community and the public alike were what many would regard as radical suggestions, such as car-free neighbourhoods and community kitchens, but also images of an enjoyable daily life with short distances to everyday activities, attractive green spaces and accessible public spaces for all age groups.

The images of the future Brøset presented in the design proposals became an important basis not only for the development of the zoning plan but perhaps even more for public debate on what sustainable communities may be. The issue of reputation and image building did, however, also prove to be critical in other respects. From being described and discussed in mainly positive terms – such as 'green', 'attractive' and 'future oriented' – public discourse on the Brøset project changed during the process as it came to be described as somewhat unrealistic and an area for only special interest idealists. Tensions between concrete manifestations of a three-ton society and the open-endedness and

Photos 5.4 and 5.5 Images from design proposals for the Brøset neighbourhood.
Above: Team CODE (2011). Below: Team SLA (2011).

Source: © Trondheim Municipality.

unpredictability of long-term planning processes are difficult to address without a strategy for how images of the future development are to be framed. One should not underestimate the potential conflicts between existing ways of life and what may prove necessary in order to reduce the carbon footprint of modern societies within safe levels.

Planning process – methodological pluralism and capacity building

Addressing emission and energy targets as well as objectives for becoming an attractive neighbourhood with a high quality of life requires moving into largely unknown territory and may pose considerable challenges. In the Brøset project, a combination of various quantitative and qualitative methods was used to provide the best possible knowledge basis for the choices made: focus group interviews with potential future residents; space syntax analysis; greenhouse gas emissions accounting; life-cycle costs; secondments; and design activism. In addition, there was intensive collaboration between researchers and planners, who exchanged and developed knowledge. The neighbourhood plan, in contrast to many other compact city projects (Røe, 2015), was based on predictions of future residents and the social practices they were most likely to develop.

While relying solely on standardized tools may create an artificial feeling of accuracy and conciseness, the methodological innovation and pluralism that characterized the Brøset project strengthened competence and built capacity among stakeholders. This may turn out to be useful in the long run for future developments, even if the project itself is shelved.

Keeping trajectories open

New neighbourhoods should be designed to support environmentally friendly choices, but must also allow for future residents' own initiatives and creativity. This is necessary in order to create involvement and commitment but also to ensure that the neighbourhood will serve as an attractive place to live, and to allow for changes in social or physical conditions. However, there is a risk that future residents of Brøset will not act in accordance with stated intentions, and that they might 'domesticate' (Sørensen, 2005) their surroundings in ways not foreseen by those who planned and designed the area. Future users will undoubtedly, in one way or another, define their own residential practices and it is extremely difficult to predict the outcome of this complex interaction between human and non-human elements, as demonstrated throughout architectural history. Therefore, it is necessary to keep some trajectories open for new uses, interpretations and technologies that may enter at a later stage.

Conclusions

A new culture will be shaped that we believe will be self-reinforcing because the climate-friendly alternatives will be more present, while the harmful ones will be less present . . . We will build an urban neighbourhood where the combination of size and framework conditions will provide the critical mass needed to create a culture in which neighbours and family members become each other's coaches, can assist, encourage, and show each other possibilities, and together develop and determine solutions that will enable people to enjoy a carbon-neutral life.

(Team CODE, 2011: 1)

Photos 5.6 and 5.7 Images from design proposals for the Brøset neighbourhood.
Above: Team CODE (2011). Below: Team SLA (2011).

Source: © Trondheim Municipality.

The above quotation from one of the teams taking part in the commissioning process for the Brøset project illustrates how architects may regard their role as shapers of a new residential culture. Architectural history shows abundant examples of utopian urban visions that failed to achieve their goals. A reason for this may be a failure among planners and architects to recognize that users are characterized by instability and flux, as are their interactions with their material surroundings. This touches on a recurring challenge when planning for an uncertain future: how to find the best trade-off between making things tangible and keeping them open at the same time.

The sociologist Richard Sennett brings important issues into this discussion in his writings on the 'open city' and the problem of overdetermination (e.g. Sennett, 2006, 2013). The open city is, according to Sennett, incomplete, errant, conflictual and non-linear. It does not make too many assumptions about future urban life since these kinds of assumptions may favour closure, but it embraces 'less reassuring, more febrile ideas of living together, those stimulations of differences, both visual and social, which produce openness' (Sennett, 2013: 14). Open-endedness does not imply a dismissal of design. On the contrary, Sennett claims that 'the spatial forms density takes' are essential in the process of the open city (Sennett, 2013: 8). He suggests three ways in which an open city can be well designed. First, instead of boundaries and walls, there should be 'ambiguous edges' between parts of the city, porous borders or 'passage territories' where people can interact: 'Zones in which different communities of a city meet' (Sennett, 2006: n.p.). Second, architects should aim to create 'incomplete form' in buildings and their surroundings – architecture that can be supplemented or revised instead of replaced. Buildings should be living, evolving structures, not 'fit-for-one-purpose structures' (Sennett, 2006: n.p.). Finally, urban planning should support 'unresolved narratives' of development. It should be seen as exploring the unforeseen and the role of planners thus to shape processes of exploration: 'Open-city planning attends to conflicts and possibilities in sequence; there's problem-solving, but also problem-finding, discovery rather than merely clarity' (Sennett, 2013:14).

The main driver of more sustainable residential practices is *quality*. Architects and planners must provide convincing images and give form not only to compact but also to high-quality residential solutions. Quality requires not only talented architects and planners, but capable developers, an innovative construction industry and visionary as well as dynamic politicians and planning authorities. There must be a will to use policy instruments that makes it possible for small businesses and shops to survive in urban centres integrated in residential neighbourhoods, that ensure social diversity, safeguard spaces for recreation and public life, and that makes it easy and comfortable to move around without the use of private cars.

Transformation towards low-carbon societies and sustainable communities requires a radical reconfiguration of values and consumption patterns, which are closely connected to residential practices. In the Norwegian context, this implies moving from the dream of a detached house with a private garden towards homes which are closely integrated in vital, connected, compact and open cities. Attractive urban areas and innovative architecture is not enough, nor is densification. Unless architects and planners are willing to push the limit of what developers and real estate agents consider the market wants, or what is economically feasible, the required changes in residential practices will not occur. The development of sustainable communities will also require substantial efforts from all of the relevant stakeholders in urban development.

References

Andersen, A. (2004) 'Boliger, omgivelser og miljø', *Samfunnspeilet*, 4/2004. Oslo/Kongsvinger: Statistics Norway.

Aune, M. and Støa, E. (2016) 'Second homes in Norway: multiple homes and patterns in distributed ways of residing', in Gromark, S., Ilmonen, M., Paadam, K. and Støa, E. (eds) *Ways of Residing in Transformation: Interdisciplinary Perspectives*. London: Ashgate, pp. 57–75.

Bøeng, A. C., Isaksen, E., Jama, S. M. and Stalund, M. (2011) *Energiindikatorer for Norge 1990–2009*. SSB-rapporter 31/2011. Oslo/Kongsvinger: Statistisk Sentralbyrå.

Bramley, G., Dempsey, N., Power, S., Brown, C. and Watkins, D. (2009) 'Social sustainability and urban form: evidence from five British cities', *Environment and Planning A*, 41: 2125–2142.

By- og Boligministeriet (2001) *Økologi og adferd i udvalgte boligområder*. Copenhagen: By og Boligministeriet.

Duun, H. P. (1996) 'Vi kan ikke fortette oss til bærekraftig til miljøvennlig transport', *PLAN*, 5: 50–53.

FBBB (2010) 'Om Byøkologi'. Available at: www.fbbb.dk/Hvadpercent20erpercent20by økologi, accessed 22 December 2015.

Framtidens byer (2012) 'Bedre bymiljø. Workshop program'. Available at: www. regjeringen.no/globalassets/upload/subnettsteder/framtidens_byer/kalender/2012/kort_presentasjon_av_satsingen_bedre_bymiljo_saa_langt.pdf, accessed 22 December 2015.

FutureBuilt (2015) 'What is FutureBuilt?' Available at: www.futurebuilt.no/english1, accessed 22 December 2015.

Green Building Alliance (2015) 'Grønn Byggallianse'. Available at: http://byggalliansen. no/, accessed 22 December 2015.

Grytli, E. and Støa, E. (eds) (1998) *Fra årestue til smarthus*. Oslo: Arkitekturforlaget.

Guttu, J. (2003) *Den gode boligen" Fagfolks oppfatninger av boligkvalitet gjennom 50 år*. Ph.D. thesis. Oslo: AHO.

Guttu, J. (2008) 'Profittstyrte arkitekter og handlingslammede myndigheter? Uterom i tett by – Erfaringer fra en studie i fire norske byer'. Unpublished presentation, November.

Guttu, J. and Schmidt, L. (2012) 'Nye byboliger i et kvalitetsperspektiv', in Nordahl, B. (ed.) *Boligmarked og Boligpolitikk*. Trondheim: Akademika Forlag, pp. 141–167.

Haagensen, T. (2015) *Byer og miljø – Indikatorer for miljøutviklingen i 'Framtidens byer'*. Report 2015/20. Oslo/Kongsvinger: Statistics Norway.

Hansen, H. (2012) 'Brøset er et luftslott', *Adresseavisen*, 16 February.

Hanssen, G. S., Hofstad, H. and Saglie, I.-S. (2015a) *Kompakt byutvikling. Muligheter og utfordringer*. Oslo: Universitetsforlaget.

Hanssen, G. S., Hofstad, H. and Saglie, I.-S. (2015b) 'Håndtering av motstridende hensyn i byutviklingen – tilsiktede og utilsiktede konsekvenser', in Hanssen, G. S., Hofstad, H. and Saglie, I.-S. (eds) *Kompakt byutvikling. Muligheter og utfordringer*. Oslo: Universitetsforlaget, pp. 259–270.

Hofstad, H. (2015) 'Folkehelse – vitalisering av sosial bærekraft i kompakt byutvikling', in Hanssen, G. S., Hofstad, H. and Saglie, I.-S. (eds) *Kompakt byutvikling. Muligheter og utfordringer*. Oslo: Universitetsforlaget, pp. 207–218.

Holden, E. and Norland, I. T. (2005) 'Three challenges for the compact city as a sustainable urban form: household consumption of energy and transport in eight residential areas in the Greater Oslo Region', *Urban Studies*, 42(12): 2145–2166.

Johnsen, A. B. (2014) 'Stoltenberg: Bo tett og reis kollektivt', *VG*, 16 September.

Larssæther, S., Støa, E. and Wyckmans, A. (2014) 'Future perspectives', in Støa, E., Larssæther, S. and Wyckmans, A. (eds) *Utopia Revisited: Towards a Carbon-Neutral Neighborhood at Brøset*. Bergen: Fagbokforlaget, pp. 351–355.

Müller, D. K. (2007) 'Second homes in the Nordic countries: between common heritage and exclusive commodity', introduction to special issue of *Scandinavian Journal of Hospitality and Tourism*, 7(3): 193–201.

Murphy, K. (2012) 'The social pillar of sustainable development: a literature review and framework for policy analysis', *Sustainability: Science, Practice and Policy*, 8(1): 15–29.

Næss, P., Saglie, I. L. and Thoren, K. H. (2015) 'Ideen om den kompakte byen i norsk sammenheng', in Hanssen, G. S., Hofstad, H. and Saglie, I.S. (eds) *Kompakt byutvikling. Muligheter og utfordringer*. Oslo: Universitetsforlaget, pp. 36–47.

Norwegian Environment Agency (2014) *Knowledge Base for Low-Carbon Transition in Norway – Summary*. M-287. Oslo: Norwegian Environment Agency.

Norwegian Government (2015) *State Budget*. Available at: www.regjeringen.no/no/dokumenter/Prop-1-S-2014--2015/id2005434/?ch=2, accessed 22 December 2015.

Norwegian Ministry of Climate and Environment (2014) *Proposisjon til Stortinget (forslag til stortingsvedtak)*. Prop 1S (2014–2015). Oslo: Norwegian Ministry of Climate and Environment.

Norwegian Ministry of Local Government and Modernization (2014) *Housing for Welfare: National Strategy for Housing and Support Services (2014–2020)*. H-2312 E. Oslo: Norwegian Ministry of Local Government and Modernization.

Norwegian Ministry of Local Government and Modernization (2015a) *Framtidens byer (2008–2014)*. Available at: www.regjeringen.no/no/tema/kommuner-og-regioner/by--ogstedsutvikling/framtidensbyer/id547992/, accessed 22 December 2015.

Norwegian Ministry of Local Government and Modernization (2015b) *The City as a Resource*. H-2328. Oslo: Ministry of Local Government and Modernization.

Norwegian Ministry of Local Government and Modernization (2015c) *By- og stedsutvikling. Bedre miljø i byer og tettsteder*. Available at: www.regjeringen.no/no/tema/kommuner-og-regioner/by--og-stedsutvikling/id1235/, accessed 13 October 2015.

Norwegian Ministry of Local Government and Regional Development (2011) *Rom for alle. En sosial boligpolitikk for framtiden*. NOU 2011:15. Oslo: Norwegian Ministry of Local Government and Regional Development.

Norwegian Ministry of Local Government and Regional Development (2013) *Byggje – bu – leve. Ein bustadpolitikk for den einskilde, samfunnet og framtidige generasjonar*. Meld. St. 17 (2012–2013). Oslo: Norwegian Ministry of Local Government and Regional Development.

Norwegian Ministry of the Environment (2000) *Miljøbyprogrammet*. T-1352. Oslo: Norwegian Ministry of the Environment.

Norwegian Ministry of the Environment (2008) *Enighet om nasjonal klimadugnad*. Available at: www.regjeringen.no/no/aktuelt/enighet-om-nasjonal-klimadugnad/id496878/, accessed 22 December 2015.

Norwegian Ministry of the Environment (2012) *Norsk Klimapolitikk*. Meld. St. 21(2011–2012), Oslo: Norwegian Ministry of the Environment.

Norwegian State Housing Bank (2014) *Program for områdeløft. Programbeskrivelse*. Oslo: Norwegian State Housing Bank.

Rambøll (2015) *Følgeevaluering av Framtidens byer. Sluttrapport*. Oslo: Rambøll Consulting Group.

Røe, P. G. (2015) 'Iscenesettelser av den kompakte byen – som visuell representasjon, arkitektur og salgsobjekt', in Hanssen, G. S., Hofstad, H. and Saglie, I.-S. (eds) *Kompakt byutvikling. Muligheter og utfordringer*. Oslo: Universitetsforlaget, pp. 48–57.

Sæther, J.-P. (2010) 'Boforhold og boligøkonomi. De fleste bor romslig i eide boliger'. *Samfunnspeilet*, 2010/5–6. Oslo/Kongsvinger: Statistics Norway.

Schmidt, L. (2015) 'Bokvalitet og sosial bærekraft', in Hanssen, G. S., Hofstad, H. and Saglie, I.-S. (eds) *Kompakt byutvikling. Muligheter og utfordringer.* Oslo: Universitetsforlaget, pp. 161–175.

Selvig, E. and Opheim, R. (2009) *Fremtidens byer – oversikt over byenes historiske utslipps utvikling og de mål for reduksjon av klimagassutslipp som er nedfelt i handlingspro grammene vinter 2009.* Oslo: Civita.

Sennett, R. (2006) 'The Open City'. Available at: https://lsecities.net/media/objects/articles/the-open-city/en-gb/, accessed 8 July 2016.

Sennett, R. (2013) 'The Open City'. Available at: www.richardsennett.com/site/senn/templates/general2.aspx?pageid=38&cc=gb/, accessed 8 July 2016.

Shove, E., Watson, M., Hand, M. and Ingram, J. (2007) *The Design of Everyday Life.* Oxford: Berg.

Skjeggedal, T. (1996) 'Fortettingsoptimisme og CO2-moralisme. Om miljøvennlig tettsteds utvikling', *PLAN*, 1–2: 70–81.

Solli, C. and Bohne, R. A. (2014) 'Carbon-neutral settlements – the operationalization of a vision', in Støa, E., Larssæther, S. and Wyckmans, A. (eds) *Utopia Revisited: Towards a Carbon-Neutral Neighborhood at Brøset.* Bergen: Fagbokforlaget, pp. 89–107.

Sørensen, K. H. (2005) 'Domestication: the enactment of technology', in Berker, T. Hartmann, M., Punie, Y. and Ward, K. (eds) *Domestication of Media and Technology.* Maidenhead: Open University Press, pp. 40–61.

Sørvold, J. (2011) 'Den boligpolitiske vendingen'. Attachment to Norwegian Ministry of Local Government and Regional Development (2011) *Rom for Alle.* NOU 2011:15. Oslo: Norwegian Ministry of Local Government and Regional Development.

Statistics Norway (2002a) 'Folke- og boligtellingen. Innvandrernes boforhold, 2001'. Available at: www.ssb.no/fobinnvbolig/, accessed 21 December 2015.

Statistics Norway (2002b) 'Folke- og boligtellingen. Studenters bosted og boforhold, 2001'. Available at: www.ssb.no/befolkning/statistikker/fobstud/hvert-10-aar/2002-11-21, accessed 21 December 2015.

Statistics Norway (2011) 'Continued growth in urban settlements'. Available at: www.ssb.no/english/subjects/02/01/10/beftett_en/, accessed 21 December 2015.

Statistics Norway (2012) 'Population and housing census, households, 2011'. Available at: www.ssb.no/en/befolkning/statistikker/fobhushold/hvert-10-aar/2012-12-18, accessed 7 March 2016.

Statistics Norway (2013a) 'Økning i transportens energibruk og klimagassutslipp'. Available at: www.ssb.no/natur-og-miljo/artikler-og-publikasjoner/okning-i-transportens-energibruk-og-klimagassutslipp, accessed 17 December 2015.

Statistics Norway (2013b) 'Boforhold, levekårsundersøkelsen, 2012'. Available at: www.ssb.no/bygg-bolig-og-eiendom/statistikker/bo/hvert-3-aar/2013-02-05#content, accessed 21 December 2015.

Statistics Norway (2014) 'Utvikling i inntektsulikhet'. Available at: www.ssb.no/natur-og-miljo/barekraft/utvikling-i-inntektsulikhet, accessed 14 December 2015.

Statistics Norway, (2015a) *This is Norway 2015.* Oslo/Kongsvinger: Statistics Norway.

Statistics Norway (2015b) 'Labour force survey, Q3 2015'. Available at: www.ssb.no/en/aku/ and www.ssb.no/217366/arbeidsledige-aku-etter-kjonn-og-alder.arsgjennomsnitt.1-000-og-prosent-sa-220, accessed 17 December 2015.

Statistics Norway (2015c) *Økonomiske analyser* 3/2015. Oslo: Statistics Norway.

Statistics Norway (2015d) 'Population, 1 January 2016, estimated'. Available at: www.ssb. no/en/befolkning/statistikker/folkemengde/aar-berekna, accessed 4 January 2016.

Statistics Norway (2015e) 'Population and land area in urban settlements, 1 January 2015'. Available at: www.ssb.no/en/befolkning/statistikker/beftett, accessed 17 December 2015.

Statistics Norway (2015f) 'Emissions of greenhouse gases, 2014, preliminary figures'. Available at: www.ssb.no/en/natur-og-miljo/statistikker/klimagassn, accessed 17 December 2015.

Statistics Norway (2016) 'Immigrants and Norwegian-born to immigrant parents, 1 January 2016'. Available at: www.ssb.no/en/befolkning/statistikker/innvbef/aar/2016-03-03, accessed 7 March 2016.

Støa, E. (1996) *Boliger og kultur. Norske boligfelt på åtti-tallet sett i lys av beboernes boligidealer.* Doctoral thesis. Trondheim: Institutt for Byggekunst, Norges tekniske høgskole.

Støa, E. (2009) 'Housing in the sustainable city', in Holt-Jensen, A. and Pollock, E. (eds) *Urban Sustainability and Governance: New Challenges in Nordic-Baltic Housing Policies.* Oslo: NOVA, pp. 31–48.

Støa, E. (2012) 'Adaptability', in Smith, S. J. *et al.* (eds) *The International Encyclopedia of Housing and Homes.* Oxford: Elsevier, pp. 51–57.

Støa, E. (2014) 'Balancing social and environmental goals', in Støa, E., Larssæther, S. and Wyckmans, A. (eds) *Utopia Revisited: Towards a Carbon-Neutral Neighborhood at Brøset.* Bergen: Fagbokforlaget, pp. 109–127.

Støa, E. and Aune, M. (2012) 'Sustainable housing cultures', in Smith, S. J. *et al.* (eds) *The International Encyclopedia of Housing and Homes.* Oxford: Elsevier, pp. 111–116.

Støa, E., Høyland, K. and Wågø, S. (2006) *Bokvalitet i små boliger. Studier av fem boligprosjekter i Trondheim.* SINTEF Byggforsk Report No. 51 A06004. Trondheim: SINTEF Byggforsk.

Støa, E., Larssæther, S. and Wyckmans, A. (eds) (2014) *Utopia Revisited: Towards a Carbon-Neutral Neighborhood at Brøset.* Bergen: Fagbokforlaget.

Støa, E. and Larssæther, S. (2014) 'New house – new lifestyle? Architecture as a part of sustainable transitions at Brøset', in Støa, E., Larssæther, S. and Wyckmans, A. (eds) *Utopia Revisited: Towards a Carbon-Neutral Neighborhood at Brøset.* Bergen: Fagbokforlaget, pp. 193–213.

Støa, E. and Manum, B. (2013) *Friluftsliv og bærekraft: en del av problemet, eller en del av løsningen? Dokumentasjonsrapport case 2: Fritidsboliger.* Trondheim: NTNU.

Svane, Ø. and Wijkmark, J. (2002) *När ekobyn kom till stan. Lärdomar från Ekoporten och Understenshöjden.* Stockholm: Formas.

Team CODE (2011) *Tre tonn Brøset.* Design proposal for the Brøset area. Available at: www.trondheim.kommune.no/content/1117730295/Parallelloppdraget, accessed 4 January 2016.

TNS Gallup (2015) 'TNS Gallup Klimabarometer 2015'. Available at: www.tns-gallup.no/sokeresultat?q=klimabarometeret, accessed 21 December 2015.

Trondheim City Planning Office (2010) *Planprogram Brøset – En klimanøytral bydel.* Trondheim Municipality: City Planning Office.

Trondheim City Planning Office (2013) *Områdeplan for Brøset.* Trondheim Municipality: City Planning Office.

Vallance, S., Perkins, H. C. and Dixon. J. E. (2011) 'What is social sustainability? A clarification of concepts', *Geoforum*, 42: 342–348.

Vestergaard, H. and Haagerup, C. (2016) 'From building to boost – changing housing policies and ways of residing', in Gromark, S., Ilmonen, M., Paadam, K. and Støa, E. (eds) *Ways of Residing in Transformation: Interdisciplinary Perspectives*. London: Ashgate, pp. 21–32.

WCED (1987) *Our Common Future*. Oxford: Oxford University Press.

Winston, N. (2013) 'Sustainable communities? A comparative perspective on urban housing in the European Union', *European Planning Studies*, 22(7): 1384–1406.

ZEB (2015) 'The Research Centre on Zero Emission Buildings'. Available at: www.zeb. no/, accessed 22 December 2015.

6 Denmark

Jacob Norvig Larsen, Toke Haunstrup Christensen, Lars A. Engberg, Freja Friis, Kirsten Gram-Hanssen, Jesper Rohr Hansen, Jesper Ole Jensen and Line Valdorff Madsen

Introduction

Denmark can in many ways be considered a frontrunner as regards sustainable communities and housing. The housing sector provides high-quality housing for the vast majority of the population and in relation to the environmental perspective of sustainability the Danish Building Regulations ensures that all new buildings have a high energy efficiency standard. Further, the energy supply system has changed since the 1990s to contain a substantial input from renewable energy sources. Denmark is also often mentioned as a frontrunner regarding innovation and regulation within fields such as energy efficiency in buildings, energy retrofitting and smart grid, and scores high when deploying an elaborated approach to sustainable communities (Winston, 2013). However, Denmark is at the same time among the countries with the highest CO2 emissions per capita: in 2013, it ranked 44th out of 208 countries on this indicator (EDGAR, 2015). With a 2013 emission of 7.4 tonnes per capita, Denmark is on a par with the EU28 average, but well above the global average of 5.0. The level of comfort and the number of square metres per person of housing are continuously increasing, which makes the rebound effect a main explanation of why 'frontrunning' by itself is not an adequate condition for achieving sustainable housing and communities (Gram-Hanssen, 2013).

In this chapter, we delve into the tension between frontrunner and major consumer. We do so by exploring how four different cases of sustainable communities and housing contribute to this schism and by discussing sustainability parameters as well as probable answers on different levels of planning policy. Based on this overview of policy and indicators, the last section introduces cases that in an innovative fashion point out future development trajectories for sustainable communities and housing, and initiate a discussion concerning the role of public regulation versus civil society initiatives.

Denmark's population is 5.6 million, and with 2.9 million households this results in an average family size as low as 1.9 persons (Statistics Denmark, 2012). The country has witnessed decreasing birth rates and a general ageing of the population, but due to immigration the population has still increased slightly. The overall labour market participation rate is 74 per cent (male 76 per cent, female 72 per cent) for the population between 16 and 64 years, which is among the highest in Europe (Statistics Denmark, 2015a). Ageing of the population in combination with low birth rates and difficulty in keeping labour market participation high have resulted in the emergence of a policy discourse which stresses the future pressure on public budgets. Accordingly, the policy response has recently been to limit access to early retirement and to implement cuts in social security and unemployment benefits. The unemployment rate was 2.6 per cent in 2008, 6.2 in 2010, and 4.5 per cent in 2015.

Denmark is a capitalist political economy and the welfare state regime follows the Scandinavian social democratic model, with principles of universalism and social rights extended to both the working and the middle classes. This entails the pursuit of high equality and high-quality service provision for all (Esping-Andersen, 1990). The regional government level is almost wholly concerned with the provision of hospital services, while state and local levels address everything else. Thus, responsibilities for housing policy, urban and regional planning, environmental protection and related matters are mainly allocated at the state and local government levels. At national level the Planning Act regulates developments in urban as well as rural areas, and responsibility for housing policies, with related sustainability matters divided between the following ministries:

- Climate, Energy and Building: climate and energy-related policies in general as well as climate and energy-related issues related to construction, including all issues related to power supply.
- Environment: environmental protection and planning policy related to regulations concerning urban planning; nature preservation and nature-based tourism and experiences.
- Housing, Urban and Rural Affairs: urban and housing policies as well as a balanced rural–urban development.

Most of the administrative responsibility regarding planning and environmental protection has been decentralised to the municipalities. In relation to housing policy and related sustainability matters, regional authorities administer or run:

- Act of Business Support by facilitating Growth Forums involving regional stakeholders.
- Regional Knowledge Centres related to environment and resources.
- Facilitates and supports the formation of Local Agenda 21 strategies.

At the municipal level, the planning of housing, strategic sustainable development and transport is implemented, with the exception of regional and national infrastructure planning. Many municipalities have made voluntary agreements with state agencies, NGOs and EU bodies in order to achieve specific energy-saving objectives and strategies related to housing issues, such as:

- Strategic Energy Planning, supported by the Danish Energy Agency. Facilitates collaborative initiatives between municipalities, energy companies and local businesses. This initiative aims to spur green growth and includes advisory services and minor grants.

- Partnership agreement between the Association of Local Governments in Denmark and the Danish Energy Agency in order to promote the transformation of the municipal power supply.
- Climate Municipality Agreement with the Danish Society for Nature Conservation, entailing an annual 2 per cent reduction of carbon emissions.
- A number of Danish cities have joined the Covenant of Mayors, a pan-European association which encourages its member cities to achieve 20 per cent reduction in carbon emissions by 2020.

The housing system in Denmark

About half of the Danish population live in owner-occupied homes, typically single-family houses. The Danish housing market is influenced by urbanisation and geographically uneven economic growth. It is also heavily influenced by population concentration in the major conurbations around Copenhagen and Aarhus, and housing prices are markedly higher in the large cities compared with surrounding municipalities. In some peripheral municipalities price levels can drop to a fifth of those in the capital (Larsen *et al.*, 2014). As a result of the global financial meltdown in 2007–2008, the Danish housing market (peaking in the third quarter of 2007) experienced an unprecedented decline over seven quarters through to 2009. The economic crisis exacerbated the geographical imbalance of the housing market. Though average real house prices stabilised from 2009 to 2011 (Lunde, 2012), up to 2013 there was a continuing decline in single-family house prices throughout most of the country, except for the capital region, which experienced an average 18 per cent price increase (Larsen *et al.*, 2014).

Cooperative housing is a tenure form between private ownership and rental. Residents have collective ownership of the property and acquire the right to live in an apartment when buying a share in the building and becoming a member of the cooperative. Residents elect a management committee run by residents in charge of asset management decisions. Before 2005, cooperative housing was priced below market prices, but the system has been deregulated, and most cooperative housing is now traded on market terms, making the tenure form quite similar to owner-occupied housing.

The social housing sector accounts for about 20 per cent of the total housing stock, with 8 per cent housing cooperatives and 17 per cent private renters. In 2013, the average number of square metres per occupant was 52; homeowners occupied on average 56 square metres and renters 46. From 2011 to 2015, the number of people living in rented housing rose by 7 7 per cent, and the number of people living in owner-occupied housing decreased by 0.6 per cent (Statistics Denmark, 2015b). There are several reasons for this development: more young people are renters; an increasing number of single person households primarily rent their homes; non-Western immigrants and their descendants primarily live in rental housing. Due to the economic crisis, there has been a decline in housing

Table 6.1 Development of the Danish housing market

	1990 %	1998 %	2005 %	2013 %
Owner-occupied dwellings	52	51	49	47
Private rental housing	18	19	17	17
Social housing	17	19	21	20
Cooperative housing	4	6	7	8
Other	9	5	5	9
Number of housing units (,000)	2,353	2,461	2,601	2,762

Sources: Ministry of Urban and Housing Affairs (1998); Ministry of Social Affairs (2006); Statistics Denmark (2014)

construction with a relative increase in social housing units compared with privately owned housing units.

The Danish rental system is similar to what Kemeny (2006) describes as an 'integrated rental market model'. Social housing is characterised by universal access, a history of social democratic corporatism and a non-profit social renting sector of a size similar to that of the for-profit renting sector. The social housing sector is managed by private, non-profit housing associations governed by a system of democratically elected tenants. Social housing rents are subsidised and based on the principle of cost-price renting. The cost of social housing units should not exceed building expenses minus public subsidies in combination with the financial and administrative costs of running the estates. Profits are earmarked for building and maintenance purposes. Each housing section is a financially independent unit and no cross-subsidy or pooling of costs between housing sections is allowed. When mortgages are amortised, rent contributions continue and finance two fund systems: a local fund at the disposal of the individual housing association as a financial buffer and working capital in relation to local problems; and the National Building Fund that finances major repair works, rent reductions and social investments in the sector.

Private rental housing includes old, cheap apartments, new, expensive apartments, and everything in between these two extremes. Quality ranges from poor to excellent, but price levels do not necessarily reflect quality. Thus, for dwellings established before 1992, the rent is regulated by law and reflects the cost of operating the building. For newer dwellings, the rent level is mainly market-based. However, according to existing regulation, property owners are allowed to increase rent levels provided refurbishments are made. In a number of cases such refurbishments are little more than window dressing, resulting in higher rent levels in return for only marginally improved dwelling quality.

A comparison of the housing conditions of the poorest sector of the population (the lowest income quartile) in fifteen EU countries showed that Danish low-income households have above-average housing standards. A larger proportion of people in this group are homeowners, and the housing units are generally larger than in comparable groups in other countries. Also, low-income

Danish households normally have no difficulty paying their heating bills (Kristensen, 2004).

In social housing, the National Building Fund system plays a key role in improving the social sustainability of distressed neighbourhoods (which are primarily social housing districts), funding housing renovations, rent reductions and social outreach programmes and preventive measures. Between 1991 and 2016, the fund distributed approximately €6.7 billion for housing renovations, and between 2007 and 2014 €0.6 billion for social measures and rent reductions (Ministry of Housing, Urban and Rural Affairs, 2015a). As regards energy savings in the social housing sector, however, it seems that strict regulation may be causing increased rather than lowered energy consumption. For example, according to the Danish Building Regulations, renovation should improve accessibility for people who are physically disabled, which in practical terms means that it is mandatory to install elevators. Unfortunately, this, in turn, leads to higher power consumption.

In private housing, the Danish Act on Urban Renewal and Urban Development plays a key role in the housing and urban policies of Danish local governments. The law subsidises targeted urban regeneration in four areas: building renovation, condemnation, improvement of recreational areas and integrated area-based regeneration. Since 1990, approximately 74,000 private dwellings have been improved by state-subsidised building renovation, and the number of dwellings lacking one or more installations (toilet, bath or central heating) dropped from 324,000 to 120,000 between 1990 and 2015 (Ministry of Housing, Urban and Rural Affairs, 2015b).

Sustainable communities and housing in Denmark

Sustainability policy and indicators

Both state- and local-level policies and indicators for sustainability have been developed. Following the Brundtland Report of 1987, from the 1990s until the beginning of the new century Denmark witnessed a high level of activity at state level in terms of definition and work with sustainability policy and indicators combining social, economic and environmental factors. In 2001, the Danish government published the first national strategy for sustainable development together with a set of indicators, and four years later the Ministry of Housing published an action plan for 'Urban Ecology' (*Byøkologi*), as a recommendation of how to work with Local Agenda 21 activities in housing and communities. Over subsequent years, these types of activities have, however, not been given the same priority. With the governmental shift in 2001 (from a government led by the Social Democratic Party to one led by liberal-conservative parties), many activities and funding programmes that targeted citizens and Local Agenda 21 activities were closed. In 2015, the government published a plan for sustainable cities with a focus on both social and green sustainability, although not much activity followed (Ministry of Housing, Urban and Rural Affairs, 2015c).

Thus, state-supported strategies aimed at local-scale citizen involvement are much less in focus today compared with the 1990s. Since that decade, it is especially the energy and climate topic which has been on the national agenda for sustainability and housing, and in 2014 a Climate Law was passed in parliament with the long-term objective that Danish energy consumption should be entirely based on renewable energy resources by 2050. Thus, the sustainability agenda on the national level has turned more towards the reduction of CO_2 emissions and energy, while the broader sustainability agenda does not receive much attention.

Consequently, present national policies regarding sustainable communities and housing are quite focused on the energy and climate aspects, and initiatives focus on more renewables in the production and on reducing energy consumption in housing.

National policies concerning energy in buildings

Since 1961, the Danish Building Regulations have included energy requirements for new buildings. These were tightened in 1972, 1978, 1995, and since 2006 have followed the EU Directive on the Energy Performance of Buildings (EPBD). They have also included provisions on the renovation of existing buildings. Much policy focus is targeted at the energy performance certification scheme (EPC), which in Denmark dates back to the 1980s and since 2006 has followed the EPBD. The EPC is mandatory for all houses sold or rented, including new buildings. As in the rest of EU, it includes a rating from A1 to G based on the calculated energy consumption of the house. In Denmark, as in some other EU member states, the EPC also includes recommendations for improvements and the rating that could be achieved if the house were renovated according to these recommendations. Evaluations of the experiences and effectiveness of EPC are, however, not encouraging as the recommendations in the EPC seem to promote improved energy efficiency only within new buildings, not in the existing building stock (Christensen *et al.*, 2014).

Since 1996, energy utilities have been legally obligated to promote energy savings under Danish law. They are free to choose their methods, typically including different types of advice, communication and economic incentives. The energy authorities require documentation from the utilities to prove that they achieve their energy-saving targets. In addition to the utilities' saving campaigns and the EPC, there are other initiatives in Denmark that use information as a means to promote energy savings; especially relevant in this respect is the Knowledge Centre for Energy Conservation in Buildings. The purpose of the centre is to collect knowledge on how to reduce energy consumption in buildings and communicate this to professional actors within the building sector, including craftsmen.

Economic incentives for households have formed only a minor part of Danish energy policy, although some examples do exist. These include a governmental Growth Fund with €200 million to stimulate the Danish construction sector in

2009 by means of grants to spur individual households' demand. The fund provided subsidies for refurbishments and building projects in private housing, including energy retrofitting renovations.

Municipal policies and activities: the strategic-quantitative 'turn'

In a historical perspective, the role of the municipalities has changed in relation to sustainable communities. Earlier, there was a considerable number of citizen-oriented activities at the municipal level, typically supported and facilitated by (Agenda 21) Green Guides employed at the local municipality or local organisations and financially subsidised by the state (Læssøe, 2001). After 2001, many of these activities closed down. Following this, activities at the municipal level in general geared down and became more strategic as focus shifted from supporting specific activities to formulating strategies for sustainable housing, local sustainable development, climate goals, facilitating and supporting local initiatives for sustainable housing.

While the Brundtland Report defined sustainability in relatively vague terms, recent national and local-sector policies apply strictly quantifiable environmental performance goals and indicators in combination with a New Public Management agenda. Consequently, public access to sustainability goals and indicators has increased, but at the cost that policies and initiatives for local sustainability that are not able to present a set of indicators and quantifiable measures fail to achieve political support. Jensen *et al.* (2014) argue that the focus on narrow climate goals and CO2 reductions has diverted attention from sustainable communities – that is, a more locally focused approach with a broader understanding of sustainability. Thereby, the conceptualisation of 'sustainable communities' becomes less tangible and more dispersed, albeit principally more measurable at a municipal level, and it is yet more difficult to identify sustainable communities when the focus is on only environmental aspects while social and economic dimensions are excluded.

The consequence of the quantitative approach to sustainability at the municipal level is that urban policies are linked to quantitative goals. In Copenhagen, for example, the annual environmental accounting includes climate and energy, green growth, drinking water, resources and waste, air quality, traffic, noise, urban nature and the knowledge and interests of the citizens. All of these themes are presented with goals and progress for the specific year (Municipality of Copenhagen, 2015). Sustainable housing and communities are not included in any of these sustainability policies, however, nor in the green accounts for the municipality, presumably because they do not easily fit into the sector oriented approach. This indicates that sustainable housing and sustainable communities have smaller roles to play today at the municipal environmental policy level. Instead, sustainable housing and sustainable communities have mainly become elements in developing attractive neighbourhoods, aspects of policies for housing and urban development.

Sustainability certification schemes for buildings and neighbourhoods

Several sustainability certification schemes for buildings and neighbourhoods have been established over recent decades in order to systematise the approach to sustainability. These include certification schemes such as the Building Research Establishment Environmental Assessment Methodology (BREEAM) in the United Kingdom, Leadership in Energy and Environmental Design (LEED) in the United States, Haute Qualité Environnementale (HQE) in France and Deutsche Gesellschaft für Nachhaltiges Bauen (DGNB) in Germany (more details below). In addition to these, there are several local certification schemes and tools for assessing sustainability at the local level. One example of the latter is the so-called 'Rosetta' used by the Municipality of Copenhagen, which evaluates urban initiatives according to a number of social, economic and environmental considerations, such as urban life, social mix, resource consumption and contribution to local business and service.

The association Green Building Council Denmark, with members from ministries, municipalities and the private sector, has established the German DGNB scheme in a Danish context since 2010. The scheme has programmes for office buildings, residential buildings, hospitals and neighbourhoods. It calculates a sustainability score for each project based on forty-five criteria (environmental, social and economic issues, as well as technical quality and process quality), which leads to a potential certification (bronze, silver or gold) of the building or neighbourhood. The establishment of the scheme can be seen as a response to the domination of climate change mitigation and adoption in public policies since around 2000 at the expense of the broader definition of sustainability. Also, it is a response to the lack of definition and quantification of sustainability that has been seen as a barrier for implementing sustainability goals in a market economy. Furthermore, it follows the global trend of sustainable buildings and neighbourhoods, where an internationally recognised certification scheme is seen as a way of making buildings and neighbourhoods more attractive for international investors and users. Further, certification can also be seen as a way to establish a shared language and standard for sustainable buildings and sustainable neighbourhoods, and in this way bridge the different standards and tools that many Danish municipalities have established. However, many questions may be asked in relation to this. For instance, the cost of certification (certification fee plus costs related to documentation) are high, which means that it primarily makes sense for large-scale projects to achieve the certification. Furthermore, and more importantly, we need to see evidence of the extent to which this sustainability policy *norm* will change the sustainability *performance* of buildings and neighbourhoods. Will certificates be granted to those who have already decided to implement sustainability measures, or will they lead to changes in the design and layout of buildings (with a positive impact on the sustainability of the neighbourhoods)?

Inspired by Dowling *et al.* (2014), the different types of policy approaches can be grouped into different policy mechanisms, as shown in Table 6.2.

Table 6.2 Summary of key state policies and programmes governing sustainable communities and urban housing in Denmark

Policy mechanism	Sustainability and urban housing focus	Policy examples
Strategy/policy framework	Framing/vision	Energy Agreement (2012–2020) Conversion of Energy in Municipalities Strategy for Energy Retrofitting of Buildings Smart-Grid Strategy
Grants	Facilitating partnerships and collaboration across societal sectors at the local level Improve standards of existing building stock Initiate positive urban development in deprived neighbourhoods	SEP grant Building renovation Integrated area-based urban regeneration programmes combining housing renovation, employment schemes, etc.
Regulation/ standards	Improvement of building standards	Danish Building Regulations (2010) Act of Promotion of Renewable Energy (2015) Sustainability certification of buildings and neighbourhoods (DGNB)
Information	Voluntary carbon-emission-reduction agreements	Climate Municipalities Covenant of Mayors
Rebate/subsidy	Support home owners' production of green energy	Solar-Cell Tax Exemption
Market	Energy retrofitting of existing building stock	Danish Building Regulations: stricter energy requirements related to retrofitting of buildings Promotion of Energy Service Contracting (ESCO) Implementing regulation of EPCs Establishment of a 'knowledge centre for energy savings in buildings' 'Energy-saving agreement', requiring energy suppliers to make energy savings among end-users 'Better homes', a public–private initiative to promote energy retrofitting among private home owners ESCO in Danish Municipalities ESCO-Light

Barriers, drivers and challenges for sustainable communities in Denmark

In the main, an aim to further the development of sustainable housing and communities in Denmark is challenged by various social and economic as well as political factors. In socio-economic terms, there is a divide between, in some regions, urban shrinkage, a housing stock with dated energy standards, abandoning of homes, ageing and outmigration and, in the capital region and other growth centres, a growing demand for housing, steady increase in housing consumption, growth in property values and increase in the consumption of energy, land, water and other resources. Some of these developments are in complex ways intrinsically linked to implementation of energy and climate policies aimed at large-scale CO2 reductions, retrofitting of the existing housing stock, and so on. Centralisation of public institutions and government administration brings about a migratory pressure on the major conurbations. Sustainability policies have increasingly been reduced to general energy and climate policies narrowly focusing on large-scale utilities in the energy sector. Excessive regulation or mutually counteracting policy tools and technologies addressing household level seem to result in, for example, increased energy consumption, even when the opposite was intended. Many policies based on monetary incentives such as tax reductions, subsidy and so on do not take into account the daily practices of households and therefore result in unintended consequences. Consequently, a major challenge in approaching the goal of sustainable housing and communities is to reach out to individual households in order to encourage and stimulate profound changes in households' consumption of not just energy but also other resources. Below, we first discuss the overall performance in relation to energy and climate policy goals before exploring a number of cases of local or voluntary initiatives to encourage a more comprehensive development of sustainability in housing and communities.

As described above, the broader sustainability goals that were prevalent at the beginning of the 1990s have been largely replaced with a more narrow focus on energy and climate since 2000. Evaluated solely on the basis of these aspects, Denmark's climate target of reducing greenhouse gas emissions by 40 per cent by 2020 compared with 1990 seems to be within reach with a number of additional measures, in particular through upscaling wind-power generation (CONCITO, 2014). The energy sector emissions are rapidly declining and this is the single largest contributor to Denmark's greenhouse gas reductions. By now, the savings obtained with the least effort have been implemented, so in general more investment is required to obtain additional energy savings. There is a need for increased focus on energy efficiency in businesses and households as well as more focus on actions that ensure a stable supply of energy in an energy system with a very large share of renewable energy (over 40 per cent wind energy plus other sources) (CONCITO, 2014). This positive trend on climate goals is, however, contrasted by the fact that Denmark is still well above the world average in CO2 emissions per capita and, if measured by the ecological footprint per capita, is among the twenty-five highest-ranking countries (Global Footprint

Network, 2015). The 2014 CO2 emission per capita was 7.4 tonnes (Danish Energy Agency, 2015); to reach the Danish goal of a sustainable low-carbon society by 2050, emissions will need to be reduced to less than 2 tonnes per capita. For Denmark, the embodied energy and resource consumption related to the import of manufactured products plays an especially important role.

There is thus a tension between the climate-friendly energy approach of Danish policy and overall consumption, implying that Denmark is still some distance from ecological sustainability.

However, two other trends within the Danish sustainable communities and housing field are presently evolving. First, there is an increasing number of NGO and citizen-driven initiatives for local transition and sustainable consumption/living. At the same time, there is considerable focus on initiatives on the sustainability certification schemes of neighbourhoods and buildings, typically driven by professionals from the municipalities or developers. These two trends seem to have developed quite independently of each other.

Below, we analyse four different cases, each demonstrating an aspect of how the idea of sustainable communities is enacted and revitalised. In Table 6.3, these four cases are presented according to the main stakeholders, drivers and regulation mechanisms, as well as to how the cases relate to the sustainability question and the tension between Denmark as frontrunner and major consumer. The cases have been chosen to show a variety of what is happening in Denmark with respect to these questions.

Danish case studies on sustainable communities and housing

The four cases are introduced with a focus on stakeholders and drivers. The main purpose is to use these cases to discuss possible future directions of sustainable housing and communities, especially in relation to the tension between being a green frontrunner and being a major consumer.

ProjectZero in Sønderborg

ProjectZero was initiated by the municipality of Sønderborg, in southern Denmark, in 2007 and builds upon a public–private partnership between the municipality, a private energy company and a utility company. The overall aim of the project is to render the town of Sønderborg and the surrounding area CO2 neutral by 2029 through a variety of initiatives facilitated by a local secretariat. Aimed at reducing CO2 emissions, the initiatives focus on facilitating new business concepts, new partnerships and developing new solutions that are climate friendly and at the same time ensure an economic benefit for companies and local citizens. The initiatives are strengthened by the involvement and participation of local citizens, shops, businesses, craftspeople and others.

According to ProjectZero's own calculations in 2013, a CO2 reduction of 23 per cent had been accomplished since the start of the project in 2007, which was close to the aim of a 25 per cent reduction in 2015 compared with 2007

Table 6.3 Summary of selected cases with a focus on five main characteristics

Cases	Stakeholders	Regulation mechanisms	Drivers	Sustainability parameters	Tensions
ProjectZero Sønderborg	Local government Energy company Private green tech companies Construction sector Private housing Citizens Others	New public governance mechanisms Legal framework for energy companies Planning framework Building Regulations National funds	Regional energy strategy Public–private partnership Private funding of energy efficiency innovation Free-of-charge energy certificates Training of craftspeople Growth/job creation	CO2 neutral 2029 Green growth Energy efficiency in buildings Energy efficiency in energy system	Can green growth deliver in terms of fossil-free economy? Targets do not fully address ecological footprint challenges Focus on (green) growth on a city level, not sustainable everyday life
The Self-Sustaining Village	Residents' co-op Local government	Permaculture methods Planning framework	Norms and values of residents Climate change and resilience Changing everyday practices/sustainable everyday life	Ultimately no footprint Self-sustaining/local food supply Local community Recycling	Not directly applicable to mainstream urban sustainable development Small-scale holistic approach to sustainable community and housing Does not address issues of social and economic sustainability at societal level
DGNB sustainability certification of buildings and neighbourhoods	Danish Green Building Council	Internationally acknowledged sustainability certification Voluntary and market-based	Visualising qualities and increasing value of building or neighbourhood	Environmental, social and economic issues, as well as technical quality and process quality	The take-up of the certification scheme is limited Will certification lead to more sustainable buildings and neighbourhoods?
Energy efficiency in social housing renovation	Social housing associations National Housing Fund Social housing association Tenants	Law on social housing National Housing Fund system Building Regulations Tenants' democracy National housing agreements	Energy requirements in Building Regulations National Housing Fund system Social housing sector as frontrunner	Social and environmental sustainability Energy efficiency in buildings	Social equity versus environmental sustainability Mixed

(ProjectZero, 2015). Since 2007, Sønderborg has experienced a declining population and loss of jobs. Therefore, it was an additional aim to create local 'green' jobs and economic growth, and ProjectZero estimates that 800 jobs have been created by the project since its launch in 2007.

One of the most successful initiatives has been the ZERO-home programme, initiated in 2010, which aimed to engage private home owners in energy retrofitting and improving the energy standards of their homes, focusing on the EU energy certificate. Independent of households' income levels, the programme offered all interested home owners a free, impartial energy consultation (with a consultant employed in the local utility company). More than a thousand home owners from a range of income groups in the area were visited and afterwards got in contact with qualified craftsmen. At the same time the programme focused on training craftsmen to be energy advisers, involving the local technical college and vocational school. Sixty-five per cent of local craftsmen participated in the training. Furthermore, ZERO-home involved local banks and real estate agents to ensure an understanding of the cash flow and financial benefits of energy retrofitting in the real estate market, while local architects created an 'inspiration catalogue' with ideas for energy retrofitting. Quarterly surveys have shown that more than half of the home owners who were visited initiated one or more retrofitting initiatives. Furthermore, more retrofitting initiatives followed for 40 per cent of the home owners, and the surveys also show that, while economic incentives are the most important drivers for the home owners at the beginning, other factors, such as increased comfort, rise in importance during subsequent initiatives (Gram-Hanssen *et al.*, 2015).

The Self-Sustaining Village on Funen

The Self-Sustaining Village was established close to a small village in a rural area on the southern part of the island of Funen. Thus, it is not an urban project, but a community project that connects to the surrounding infrastructure of villages and settlements. It was founded in 2004 on the initiative of the residents after an agreement with the municipality. The project is a frontrunner as regards local resilience in Denmark, together with a few other initiatives that build on the Transition Town movement and permaculture principles. The village comprises a cooperative structure with nineteen households living in self-built houses based on natural and recycled building materials. It has eight hectares of land for ecological production of vegetables, fruit and hay as well as livestock farming. Furthermore, the community has a greenhouse and is self-sustaining with food for three months of the year (with plans to increase this). The permaculture design of the land ensures that farming is carried out in accordance with the natural processes of the area. The Self-Sustaining Village also works as a co-housing scheme with a basic democratic structure and common food supplies and meals.

The community works on reducing CO_2 emissions and creating a resilient local area independent of oil through everyday practices, such as food growing and recycling, and creating a sustainable local community with institutions, shops, cultural activities and jobs within cycling distance. In addition to creating new sustainable practices for themselves, the residents try to reach further and get residents in the surrounding region on board the sustainability project, for example by arranging debates and social gatherings, workshops on repairing and baking and local actions, such as planting fruit trees, inviting local residents to participate. In this way the community of the Self-Sustaining Village also works with the local government, for example to plant trees on public land, and with other local residents to create a sustainable community that encompasses more than the Self-Sustaining Village itself (Self-Sustaining Village, 2015). The aim is to create a sustainable and resilient local community comprising the surrounding villages and in this way the residents take part in local planning. The case of the Self-Sustaining Village builds on the norms and values of the residents about building a resilient community as an answer to the climate change issue. The project is based on changing everyday life practices and local structures to accommodate a sustainable daily life without a CO2 footprint. It is a small-scale, holistic project that might not be directly applicable to mainstream urban development and planning, and it does not address issues of social and economic sustainability or equity on a *societal* scale. However, such bottom-up everyday life-changing projects might push sustainability developments on a local scale that can inspire changes on municipal, regional and urban scales (Marckmann *et al.*, 2012).

Retrofitting social housing

A major initiative towards upgrading the standard of the existing housing stock in Denmark comes from retrofitting social housing. This upgrading is primarily financed through the National Building Fund. In 2007–2012, 566,000 social housing units (equating to 17 per cent of all housing units) were renovated, with a total investment of €3.6 billion. Renovation is carried out according to the Danish Building Regulations and improve the energy efficiency of the buildings, mainly due to improved insulation of facades, new windows, reduction of thermal bridges, introduction of central heating controls as well as individual meters and billing for heat consumption. Consultants and administrators in the social housing sector have claimed that heat consumption is typically reduced by 30 per cent as a consequence of renovation.

Although no thorough studies have been made, a few evaluations have tested the energy reduction of a number of social housing renovations. One of these is the study by Danielsen *et al.* (2011), which evaluated data on the consumption of heat, electricity and water before and after the renovation of four social housing estates. This showed very mixed results: for heat consumption, there were reductions ranging between 7 to 26 per cent, whereas building-specific shared electricity consumption in some cases increased quite dramatically, for example due to the installation of elevators. Similarly, with water consumption, some realised savings while others increased consumption after the renovation.

Generally, retrofitting is hampered by very complex regulations, particularly as regards access to finance. Moreover, at the level of the single estate, many tenants' associations are reluctant to engage in comprehensive retrofitting activities as they mostly consider a stable rent level their top priority. Taken together, this reduces the overall social housing demand for retrofitting. Additionally:

1. Retrofitting often includes the merging of smaller flats, or expansion of existing flats. Also, the number of residents is often reduced as a result. In total, this leads to a higher housing consumption (square metres) per resident, and typically also to higher energy consumption per resident. In the four housing estates, the energy consumption per resident grew by approximately 30 per cent as a result of the renovation.
2. Electricity-consuming equipment – for example, mechanical ventilation, elevators, surveillance systems, lighting in shared areas, and so on – was often installed as part of the renovation, which typically leads to increases in electricity consumption.

All in all, this indicates that, while renovation increases the energy efficiency of *buildings* (measured in kWh/m^2), heat consumption *per person* (measured in kWh) might increase as a result of the renovation and the improved housing standard.

A main driver behind these apparently counter-efficient initiatives is the ambition to provide social housing estates with a contemporary housing standard, as well as the ambition to attract new types of residents. The housing policy goal is to achieve a more balanced composition of residents (social mix) and to improve the often negative image of social housing estates.

Sustainability certification of neighbourhoods

Over the last five years, the German standard for sustainable buildings and neighbourhoods – DGNB (Deutsche Gesellschaft für Nachhaltiges Bauen (German Society for Sustainable Building)) – has gradually been introduced in Denmark on a voluntary basis. This includes a special standard for sustainable neighbourhoods, DGNB 'New Urban Districts', which has been adapted to a Danish context. The DGNB New Urban Districts scheme assesses each neighbourhood on forty-five indicators in five thematic areas:

*   environmental quality;
*   economic quality;
*   socio-cultural and functional quality;
*   technical quality; and
*   process quality.

The indicators include a number of different evaluation parameters, such a the amount and quality of public spaces and 'place-making' in the area, the area's contribution to the municipal economy, the involvement of local actors in the development plan, the social and functional mix in the area and many others. Each of the parameters has a different weight, leading to a total score (as a percentage) that defines the degree of sustainability.

The neighbourhoods are subsequently rewarded with a gold, silver or bronze rating. Ideally, such assessment and visualisation of area-based sustainability can provide different actors with important information about the area. For municipalities and developers, it can be a way to visualise, maximise and prioritise various sustainability issues; for investors, it can provide assurance that the area holds a certain sustainability standard, making it attractive for future investments. The certification aims to make sustainability explicit and allows for consistent benchmarking across areas, which makes it clearer to communicate what is meant when a neighbourhood development plan is described as 'sustainable'.

A non-profit association, the Danish Green Building Council, decided to use the DGNB scheme, rather than Leadership in Energy and Environmental Design (LEED) from the US, the UK's Building Research Establishment Environmental Assessment Methodology (BREEAM) or the French Haute Qualité Environnementale (HQE), because it is the most recent of the four schemes, and therefore reflects the latest European standards for sustainability assessment. More importantly, Germany's Green Building Council allows the DGNB scheme to be locally adapted to a Danish context, that is the development of a Danish DGNB standard, and such an option did not exist with either LEED or BREEAM. The adaptation of DGNB New Urban Districts in Denmark has therefore consisted of two stages. First, a pilot test of the original DGNB criteria was carried out in four development areas: Nørrestrand (Horsens), Nordhavn (Copenhagen), Carlsberg City District (Copenhagen) and Brygger Bakke (Aarhus). Second, the German DGNB criteria were adapted to the Danish context. This was carried out by a number of experts (consultants, researchers, municipal planners and so on) working on a voluntary basis in groups on different themes. It is too early to conclude what future role the DGNB New Urban Districts scheme will have in Denmark. The pilot test of the four areas provides a picture of a systematic but also rather resource-intensive tool (Jensen, 2014).

The main dilemma concerns the high degree of details in data documentation, which increases credibility and legitimacy but also requires significant resources, making the scheme costly to use. Therefore, potential users, such as municipalities and private developers, have so far been reluctant to use the certification scheme at the neighbourhood scale. A low uptake in practice is a well-known challenge, and partly relates to the voluntary character of the scheme (Sharifi and Murayama, 2014). However, the adaptation process by Danish voluntary actors might have provided a large ownership of the standard, which will potentially make the DGNB New Urban Districts scheme a condensed collection of knowledge and best practice on urban sustainability in Denmark. Thereby, it may serve not only as a certification tool, but also as a reference tool for future sustainable urban development.

Conclusions

Denmark is mentioned from time to time as a frontrunner as regards innovation and regulation within fields such as energy efficiency in buildings, energy retrofitting and smart grid, and scores high when deploying an elaborated approach to sustainable communities. Moreover, within the last twenty years the national energy supply system has changed to contain a substantial input from renewable energy sources. Still, when it comes to CO2 emissions per capita, Denmark ranks a disappointing 44th out of 208 countries, at 7.4 tonnes per capita, well above the global average of 5.0 and no better than the EU28 average. There are many explanations for this paradoxical situation. One is that housing standards are already high and constantly improving, as expressed, for example, by the average number of square metres available for each individual occupant in Danish homes. Correspondingly, the average size of household has decreased, with the current level being just 1.9 persons per household. Thus, out of a total population of 5.6 million, 1.6 million live alone. In addition to larger consumption of housing area, in existing as well as new buildings, another major factor contributing to growing energy consumption is retrofitting the existing building stock. While energy performance according to building standards constantly becomes stricter, there is a parallel development where technical standards are continuously enhanced. In several cases these cause increases in energy consumption due to the use of more automated ventilation equipment, elevators and so on. Finally, the focus of public policies for sustainability, which in the wake of the first energy crises in the 1970s covered a multiplicity of sustainability variables, has in recent years focused ever more narrowly on quantitative CO2 reductions. These policies have primarily targeted large-scale actors, such as energy utilities, with little effort made to change the behaviour of consumers, renters and home owners.

Supported by different forms of national regulations, Danish municipalities are the main authorities dealing with sustainable housing and communities in Denmark. The municipalities, however, are confronted with very different challenges depending on where they are located. Near the capital and other economic growth centres, there is a huge demand for affordable housing, whereas other municipalities have to deal with the consequences of stagnating or even non-existent housing demand. Thus, the most pressing questions of sustainability vary significantly across the country.

Overall, however, there has been a tendency to focus sustainability policies and initiatives on energy and climate issues. Therefore, the main question we raise in this chapter relates to energy and climate issues, and deals with the tension between Denmark being a frontrunner country and at the same time a major consumer. In the last part of the chapter, we introduced four cases that show a variety of actors working on defining and realising sustainable housing and communities in Denmark, and it is thus relevant to discuss how each of them relates to this tension.

The tension seems most visible in the case of retrofitting social housing. Here building regulations and new efficient technologies are expected to deliver energy savings, but in reality the retrofitting tends to entail increased energy consumption. In this case the tension should be seen within a broader sustainability perspective where social goals of improved housing standards for the less affluent together with more accessible buildings for people with physical disabilities may go some way to explaining higher energy consumption rates.

The Self-Sustaining Village can be seen as an opposite case, in which residents on their own initiative have decided to live less resource-intensive everyday lives, and thus in practice try to demonstrate another way of being a frontrunner – a practical way of solving the tension. In this case, however, we might ask whether the solution has a sufficiently broad appeal to be scaled up to society level.

In 2015, the strongest case in Denmark for delivering practical solutions on a housing and community level is probably ProjectZero in Sønderborg. As with a few other municipalities in Denmark, ProjectZero aims to combine a wide range of policy measures to attain a low-carbon future, primarily through energy retrofitting of existing buildings and changes to energy production. In practice, it has shown how ecology and economy can work hand in hand to deliver more sustainable solutions in an economically declining region of the country. Thus, the project works with a broader notion of sustainability than just energy and climate, even though these are the main drivers. Nevertheless, for the residents, the main goal of energy retrofitting houses seems to be to gain more comfort rather than to reduce energy consumption.

The final case, DGNB, is somewhat different from the others as it focuses on a tool for sustainable communities and thus an approach for how to achieve them, rather than an actual example of how it works in reality. With this case, we wanted to include the diversity and richness in ways of working towards more sustainable housing and community, as it is our conclusion that this field needs many different and parallel approaches to exchange, influence and inspire the work towards more sustainable housing and communities.

References

Christensen, T. H., Gram-Hanssen, K., de Best-Waldhober, M. and Adjei, A. (2014) 'Energy retrofits of Danish homes: is the Energy Performance Certificate useful?', *Building Research and Information*, 42(4): 489–500.

CONCITO (2014) *Annual Climate Outlook 2014 (ACO 2014)*. Copenhagen: CONCITO. Available at http://concito.dk/files/dokumenter/artikler/aco2014.pdf, accessed 11 July 2016.

Danielsen, C. B., Jensen, J. O., Kirkeby, I. M., Ginnerup, S., Clementsen, A. and Hansen, M. Ø. (2011) *Renovering af efterkrigstidens almene bebyggelser. Evaluering a to renoveringer med fokus på arkitektur, kulturarv, bæredygtighed og tilgængelighed*. SBi 2011:22. Copenhagen: Danish Building Research Institute, SBi, Aalborg University.

Danish Energy Agency (2015) *Foreløbig energistatistik 2014*. Copenhagen: Danish Energy Agency. Available at www.ens.dk/info/tal-kort/statistik-nogletal/arlig-energistatistik, accessed 11 July 2016.

Dowling, R., McGuirk, P. and Bulkeley, H. (2014) 'Retrofitting cities: local governance in Sydney, Australia', *Cities*, 38: 18–24.

Emission Database for Global Atmospheric Research (EDGAR) (2015) *CO2 Time Series 1990–2013 per Capita for World Countries*. Ispra: Emission Database for Global Atmospheric Research/Joint Research Centre and Netherlands Environmental Assessment Agency. Available at http://edgar.jrc.ec.europa.eu/overview.php?v=CO2ts_pc1990-2013&sort=des9, accessed 11 July 2016.

Esping-Anderson, G. (1990) *The Three Worlds of Welfare Capitalism*. Cambridge: Polity Press.

Global Footprint Network (2015) *Ecological Wealth of Nations*. Brussels: Global Footprint Network. Available at www.footprintnetwork.org/ecological_footprint_nations/ecological_per_capita.html, accessed 11 July 2016.

Gram-Hanssen, K. (2013) 'Efficient technologies or user behaviour, which is the more important when reducing households' energy consumption?', *Energy Efficiency*, 6: 447–457.

Gram-Hanssen, K., Friis, F., Jensen, J. O., Hansen, A. R. and Bräuner, E. V. (2015) *Renovering af danske parcelhuse – eksisterende viden og nye erfaringer*. Copenhagen: Danish Building Research Institute, SBi, Aalborg University.

Jensen, J. O. (2014) 'Sustainability certification of neighbourhoods: experience from DGNB New Urban Districts in Denmark', *Nordregio News*, 1. Available at www.nordregio.se/en/Metameny/Nordregio-News/2014/Planning-Tools-for-Urban-Sustainability/, accessed 11 July 2016.

Jensen, J. O., Jensen, O. M. and Gram-Hanssen, K. (2014) 'Transition to sustainable housing and construction', in Holm, J. (ed.) *Sustainable Transition of Housing and Construction*. Frederiksberg: Frydenlund Academic, pp. 59–81.

Kemeny, J. (2006) 'Corporatism and housing regimes', *Housing, Theory and Society*, 23(1): 1–18.

Kristensen, H. (2004) *Housing in Denmark*. Copenhagen: Centre for Housing and Welfare and Realdania Research.

Larsen, J. N., Andersen, H. T., Haldrup, K., Rhiger Hansen, A., Jacobsen, M. H. and Jensen J. O. (2014) *Boligmarkedet uden for de store byer: Analyse*. SBI 2014:05. Copenhagen: Danish Building Research Institute, SBi, Aalborg University.

Lunde, J. (2012) 'Impacts on wealth and debt of changes in the Danish Financial Framework over a housing cycle', in Colin, J., Dunse, N. and White, M. (eds) *Challenges of the Housing Economy: An International Perspective*. Hoboken, NJ: John Wiley & Sons, pp. 128–152.

Læssøe, J. (2001) *Evaluering af Grøn Guide Ordningen 1997–2000*. Copenhagen: Den Grønne Fond/Miljøstyrelsen.

Marckmann, B., Gram-Hanssen, K. and Christensen, T. H. (2012) 'Sustainable living and co-housing: evidence from a case study of eco-villages', *Built Environment*, 38(3): 413–429.

Ministry of Housing, Urban and Rural Affairs (2015a) *Fact Sheet on the Danish Social Housing Sector*. Copenhagen: Ministry of Housing, Urban and Rural Affairs.

Ministry of Housing, Urban and Rural Affairs (2015b) *Fact Sheet on Urban Renewal*. Copenhagen: Ministry of Housing, Urban and Rural Affairs.

Ministry of Housing, Urban and Rural Affairs (2015c) *Bæredygtige byer – en social og grøn bæredygtig bypolitik*. Copenhagen: Ministry of Housing, Urban and Rural Affairs.

Ministry of Social Affairs (2006) *Den almene boligsektors fremtid*. Copenhagen: Ministry of Social Affairs.

Ministry of Urban and Housing Affairs (1998) *Bygge- og boligpolitisk oversigt 1997–1998*. Copenhagen: Ministry of Urban and Housing Affairs.

Municipality of Copenhagen (2015) *Københavns miljøregnskab*. Copenhagen: Municipality of Copenhagen. Available at www.kk.dk/miljoeregnskab, accessed 11 July 2016.

ProjectZero (2015) *Om ProjektZero*. Available at www.projectzero.dk/da-DK/TopPages/Om-ProjectZero.aspx, accessed 11 July 2016.

Self-Sustaining Village (2015) *Den Selvforsynende Landsby*. Available at http://selvforsyning.dk/wordpress/, accessed 11 July 2016.

Sharifi, A. and Murayama, A. (2014) 'Neighbourhood sustainability assessment in action: cross-evaluation of three assessment systems and their cases from the US, the UK, and Japan', *Building and Environment*, 72: 243–258

Statistics Denmark (2012) *Statistiske Efterretninger – Befolkning og valg*. Copenhagen: Statistics Denmark.

Statistics Denmark (2014) *Statistical Yearbook 2014*. Copenhagen: Statistics Denmark.

Statistics Denmark (2015a) *16-64 årige offentligt forsørgede*. Copenhagen: Statistics Denmark. Available at http://dst.dk/da/Statistik/emner/offentligt-forsoergede/16-64-aarige-offentligt-forsoergede.aspx, accessed 11 July 2016.

Statistics Denmark (2015b) *Boligopgørelsen. 1. januar 2015. Levevilkår*. Copenhagen: Statistics Denmark.

Winston, N. (2013) 'Sustainable communities? A comparative perspective on urban housing in the European Union', *European Planning Studies*, 22(7): 1384–1406.

7   The Netherlands

Reinout Kleinhans

Introduction

While it is one of the smaller countries in Europe, the Netherlands has a population of nearly 17 million, of which nearly half lives in densely populated areas. With a GDP per capita of €38,300 and an approximate unemployment rate of 6.5, it is one of the more prosperous countries in the European Union (see Table 1.1). After an extended economic recession, the country is now slowly recovering; it has experienced positive economic growth since 2015.

The Netherlands has traditionally engaged in high levels of state intervention in housing and welfare policies, emphasising the importance of equal opportunities. Compared to other countries, income differences are relatively small. Using Esping-Andersen's welfare state typology, the Dutch welfare state regime has been characterised as a combination of a social democratic regime with substantial state intervention and a corporatist regime with significant decentralisation and multi-actor cooperation (Hoekstra, 2003).

In many countries, neo-liberalism and welfare state retrenchment have shifted the economies and are reshaping the ways in which citizens, public, private and third sectors interact with one another (Jessop, 2002; Brenner, 2004). Due to the economic crisis, many European states are currently implementing austerity measures and cuts in public policy, alongside longer trends of welfare retrenchment. The Netherlands is no exception to this rule. A series of budget cuts and fundamental legal and systemic reforms in healthcare (especially long-term care for elderly people), social security, housing and other fields are changing the nature of the Dutch welfare state and thus also impacting on the pursuit of sustainable housing and communities. Devolution from national government to local authorities, which came into effect on 1 January 2015, is a key feature of these reforms. Local authorities now face more responsibilities yet receive less money to address them. Moreover, several measures have significantly limited the tasks and investment capacities of housing associations. As such, local opportunities and resources to achieve sustainable housing and communities from the 'top down' have seriously decreased.

As in many other European countries, the Netherlands has an increasingly ageing population. The growing proportion of older people is generating enormous societal challenges with regard to the size of the workforce and the demand for – and funding of – pensions, healthcare and other age-related services (Gavrilov and Heuveline, 2003). For example, many local governments in the Netherlands are now implementing policies that support elderly people to remain and grow old in their own homes instead of 'moving' them into old people's homes or nursing homes (Tinker *et al.*, 2013; Dijkhoff, 2014). These policies are not only considered necessary for reducing the cost of expensive institutionalized care but are also regarded as positive for the well-being of elderly people (WHO, 2007; Wiles *et al.*, 2012).

A final important socio-demographic development is population shrinkage, which is occurring in the upper north-eastern, south-western and south-eastern (rural) parts of the Netherlands. Due to a combination of ageing and selective migration of young households, the populations in these areas are decreasing slowly but steadily. This has already caused problems in terms of maintaining facilities and services in these areas, as the economic basis of stores and (public) facilities is eroding quickly. This is a serious complication for maintaining and creating sustainable housing and communities.

The housing system in the Netherlands

The Netherlands has traditionally engaged in high levels of state intervention in housing and welfare policies, emphasising the importance of equal opportunities. Historically, preserving a balance between social and private housing has been a key issue (Boelhouwer and Priemus, 2014). Owing to the redistributive effects of the welfare state – for example, through various forms of support for disadvantaged people – income differences are relatively small. Since the 1980s, many housing responsibilities have been decentralised or delegated to local authorities or housing associations. While the national government usually provides policy principles, implementation of housing and regeneration policies is one of the main responsibilities of the local authorities, housing associations and other stakeholders. Since 1995, social housing has been managed by housing associations, private institutions that function within the public framework of the Housing Act (*Woningwet*) (Priemus, 2003). These changes have had an impact on the tenure distribution of the housing stock.

Nowadays, social housing comprises almost a third of the total Dutch housing stock, whereas owner occupation amounts to approximately 60 per cent of the stock. Since the early 1970s, the share of social housing has gradually decreased from 37 per cent to 31 per cent (see Table 7.1). Nevertheless, the share of social rented housing is higher than in any other EU country. Traditionally, the private rented sector has been relatively small in the Netherlands, and its share has decreased from 28 per cent in 1971 to 9 per cent (see Table 7.1). The allocation of social housing is subject to strict eligibility rules regarding income and household size, although several housing associations are experimenting with other forms of allocation, such as lotteries (Van Daalen and Van der Land, 2008). Despite the eligibility rules, the sector has not become residualised, as it accommodates many

Table 7.1 Housing tenure in the Netherlands since 1971

Year	Owner-occupied housing (n)	Private rented housing (n)	Social rented housing (n)	Owner-occupied housing (%)	Private rented housing (%)	Social rented housing (%)
1971	1,324,028	1,046,421	1,396,627	35	28	37
1975	1,583,471	1,053,319	1,643,951	37	25	38
1980	1,931,053	1,018,397	1,797,635	41	22	38
1985	2,253,150	981,705	2,054,466	43	19	39
1990	2,628,902	940,657	2,232,803	45	16	39
1995	2,994,128	870,451	2,327,343	48	14	38
2000	3,463,843	768,592	2,357,225	53	12	36
2005	3,861,240	693,533	2,303,946	56	10	34
2010	4,254,619	646,022	2,271,795	59	9	32
2012	4,363,195	628,055	2,274,966	60	9	31

Source: ABF Research, 2012.

households whose income has risen after moving into their dwellings (Musterd, 2014). Until very recently, income checks were performed only at the time of allocation, so tenants are not obliged to move if and when their household income rises above the threshold level of eligibility. The private rented sector displays a stronger income mix, as it accommodates both low-income and higher-income households, especially in the larger cities.

The share of owner-occupied housing has increased substantially since 1971 (see Table 7.1). Approximately 60 per cent of the population now owns their dwelling with a mortgage or loan (similar to Norway and Sweden), whereas the share of owners without a loan or mortgage (8 per cent) is very low in comparative perspective. Historically, Dutch citizens have always had ample opportunities to deduct mortgage repayments from their tax bills. This implies a huge fiscal incentive for home ownership (Boelhouwer, 2014). Simultaneously, these incentives are subject to continuous discussion about financial inequalities between renters and home owners.

The special position of housing associations has allowed them to make profits that they must reinvest in their 'core business' of providing social housing for low-income households and improving the 'liveability' of neighbourhoods where that social housing is located. This system of a revolving fund has basically guaranteed that housing associations have sufficient financial resources to carry out their tasks. Due to this position, housing associations have played a key role in Urban Renewal (*Stedelijke Vernieuwing*), a national policy that was enacted in 1997 and concluded in 2015. Over the course of this period, housing associations demolished tens of thousands of social rented dwellings, mostly in post-war neighbourhoods, and replaced them with more expensive social or private rented and owner-occupied dwellings, predominantly for middle-income groups (Kleinhans, 2012).

However, the financial position of housing associations, which were responsible for a large part of the housing production in the Netherlands, has rapidly deteriorated as a result of a number of recent developments. First, in July 2012, the Dutch parliament agreed to a revision of the Housing Act (*Woningwet*) in which the duties of housing associations were explicitly delineated, including a definition of the target group as specified by the European Commission (as of 1 January 2012 raised to a maximum annual income of €34,085) and the obligation to allot at least 90 per cent of social housing available annually to this target group (Boelhouwer and Priemus, 2014: 230). Not only is this income limit lower than previously, but the fact that the overwhelming majority of vacant housing has to be allocated to the lowest income groups has removed the possibility of allocating social housing to households on lower-middle incomes – that is, households between the lowest and middle incomes. Of course, this measure has had a negative impact on the rental incomes received by housing associations.

Second, the coalition agreement of the Rutte I Cabinet, established in 2011, included an additional general budget cut of about €15 billion. As a result, the national government has looked for new ways to generate additional income, and some of the related measures substantially affect housing associations. In the coalition agreement, possibilities were created to increase rents beyond inflation. The additional rental income for housing associations will be creamed off via a new Landlord Levy (*Verhuurdersheffing*), increasing to a level of about €1.7 billion a year in 2017 (Boelhouwer and Priemus, 2014: 228). The prime reason for establishing this levy was the need to funnel extra income into the Treasury, which had to comply with the '3 per cent rule' of the European Commission. This rule requires governments of EU member states to reduce their structural budget deficits to 3 per cent of GDP. This measure has significantly eroded the investment capacity of housing associations, an issue to which I shall return later in the chapter.

Since 2015, the Dutch economy has been recovering from the Global Financial Crisis. This recovery has been reflected by increasing pressure on the housing market, especially in the large cities. A clear indication is the rise of prices in the owner-occupied sector. Within the rented sector, accessibility and affordability are under increasing pressure for several reasons. First, both the social and private rented sector are decreasing in size, while the private rented sector is usually too expensive for low-income households. Middle-income households are increasingly experiencing accessibility problems in the owner-occupied sector due to tougher mortgage conditions set by the financial institutions, and stricter income limits set on social housing (Boelhouwer, 2014). Second, rents in the social housing sector are also increasing as a result of policies to reduce the share of affluent households in this sector by prompting them to move elsewhere (Musterd, 2014) and the need for housing associations to increase their rental income to pay the Landlord Levy. Finally, the demand for social housing is accelerating due to the very high and sudden influx of asylum-seekers from the Middle East, thus increasing the pressure on existing social rented housing. Jointly, these trends have had a highly negative impact on housing affordability among young people, especially those starting their careers.

Sustainable communities and sustainable housing

Definitions used in the Netherlands

As in so many countries, 'sustainability' has become a widely used term in the Netherlands. Despite efforts to provide a conceptual and empirical clarification, there is a plethora of definitions of the concept. Many definitions pay tribute to the famous definition of the World Commission on Environment and Development in *Our Common Future*. This report defined 'sustainable development' as development that meets 'the needs of the present without compromising the ability of future generations to meet their own needs' (WCED, 1987: 8). In this context, the three 'Ps' are often also rehearsed: people, planet and profit (or prosperity).

The discourse on sustainable communities in the Netherlands has been strongly affected by the so-called 'Bristol Accord' that was agreed at the European Union Ministerial Meeting in Bristol, UK, on 6–7 December 2005. In the accord, 'sustainable communities' are defined as:

> places where people want to live and work, now and in the future. They meet the diverse needs of existing and future residents, are sensitive to their environment, and contribute to a high quality of life. They are safe and inclusive, well planned, built and run, and offer equality of opportunity and good services for all.
>
> (ODPM, 2005: 6)

Over time, sustainability has increasingly been connected to notions of innovation in and the competitiveness of the Dutch economy. This is clear in the Sustainability Agenda (*Agenda Duurzaamheid*) that was issued by the national government in 2011. This agenda states:

> The cabinet aims for growth that does not deplete the natural capital of our earth, and which strengthens our economy. Green growth also ensures that future generations can meet their needs; it opens up opportunities for Dutch trade and industry, which compete at the top of the world league in the sectors of water and food. Investing in green growth also means investing in competitive power.
>
> (Ministry of Infrastructure and the Environment, 2011: 1; my translation)

This definition is clearly based on the WCED definition, but it adds an inextricable connection with the competitive power of the Dutch economy. The term 'green growth' refers to a combination of economic growth with a decrease in pressure on natural resources (CBS, 2014).

Relevant policies

A number of Dutch policies have focused on sustainable communities and sustainable housing. The national government sets the key principles in a framework under which local authorities, housing associations, developers and other actors have to develop local policies and plans. Some policies use the term 'sustainable communities' in general terms when discussing the outcomes, while others have it as (one of) their main objective(s). An example is Urban Renewal, which was in place from 1997 until 2015. Van Bergeijk *et al.* (2008) have shown that this programme perfectly equates to the definition of 'sustainable communities' in the Bristol Accord by focusing on creating opportunities for residents and creating neighbourhoods where people want to live on positive grounds (Gruis *et al.*, 2006). However, a series of research projects has shown that this ambition has been achieved only partially and selectively, with good results from the physical part of the renewal and disappointing results with regard to outcomes for social cohesion, safety and reputation. (For an overview of policy outcomes, see Kleinhans (2012).) Moreover, the official national Urban Renewal policy concluded in 2015, resulting in an abrupt end to state funding that had amounted to millions of euros for the policy since its inception in 1997.

More recently, Dutch housing and community policies have been strongly affected by a range of policies in the fields of construction, energy and climate change. In 2008, the EU's Energy Performance of Buildings Directive (EPBD) was implemented in the Netherlands. Under this directive, all member states were obliged to establish and apply minimum energy performance requirements for new and existing buildings, ensure the certification of building energy performance and require the regular inspection of boilers and air-conditioning systems in buildings (Filippidou *et al.*, 2014).

The aforementioned Agenda Sustainability (Ministry of Infrastructure and the Environment, 2011) outlines a number of priorities: materials and production chains; sustainable water and land use; food; climate and energy; and mobility. From a distance, it appears that the emphasis on sustainability is prominent in policies relating to climate change, CO2-reduction and sustainable energy. A clear example relates to policies for saving energy. As with many other member states, the Netherlands has worked hard to elaborate EU directives on energy-saving measures that have to be operationalised at member-state level, such as compulsory environmental impact assessments of materials for new construction of housing. A consistent objective has been that these policies should result in more sustainable housing construction and renovation practices, and communities that are more sustainable in terms of energy production and use, water recycling and the availability of (public) transport options.

The focus on sustainable energy is exemplified by a national Energy Agreement for Sustainable Growth (*Nationaal Energie Akkoord*) that was forged in 2013. This agreement unites divergent interests and brings together more than forty organisations – including central, regional and local government, employers' associations and unions, nature conservation and environmental organisations, and other civil-society organisations and financial institutions. It is based on an understanding that a long-term perspective means placing the common good far above the separate interests of either individuals or organisations and that it also means a growth path defined by energy and climate objectives as well as by feasible and necessary gains in competitiveness, employment and exports (SER, 2013). Again, this reveals an inextricable connection between sustainability and relative competitiveness.

In terms of spatial planning and infrastructure, which are very important for the development of sustainable housing and communities, a significant step will be provided by the new Environment and Planning Act (*Omgevingswet*). This new law aims to integrate all current environmental legislation – dozens of laws and hundreds of regulations for land use, residential areas, infrastructure, the environment, nature and water. The Environment and Planning Act explicitly aims for sustainable development by achieving and maintaining a safe and healthy physical living environment and a high environmental quality, as well as effectively developing and maintaining the physical living environment in order to maximise its societal functions (House of Representatives, 2013). At the time of writing both the House of Representatives and the Senate still had to approve the Act, but it is expected to come into force in 2018.

In the preceding paragraphs, we have seen that the recent focus in the use of sustainability is mostly on the built environment, 'green growth', climate change and energy policies. One could argue that a social perspective on the notion of sustainable communities has gradually shifted to the background, despite the attention devoted to it in the Bristol Accord. However, it turns out that this perspective is actually still applied, albeit rarely under a '(social) sustainability' heading. Rather, the concept of sustainability is framed within a view that it is necessary to increase the contributions from, and initiatives of, citizens with regard to solving societal problems. In terms of policy discourse, the key terms are 'active citizenship', 'self-organisation' and 'do-it-yourself-democracy' (see, e.g., Boonstra and Boelens, 2011; Verhoeven and Tonkens, 2013: Kleinhans *et al.*, 2013). The Dutch Ministry of the Interior published the white paper *Do-It-Yourself Democracy* in 2013, stating that:

> The Cabinet aims to offer room and trust for societal initiatives and actively support the transition towards a do-it-yourself democracy (which is a form of citizens taking part in deciding to take up societal issues themselves) . . . Several societal trends require the Cabinet's view on this matter: a) an increasing level of self-organisation in society, b) a retrenching government, and c) an increasing demand for social connectedness. Apart from these trends, the transition to more do-it-yourself democracy is relevant from a governmental point of view, due to scaling-up, decentralisation and budget cuts.
>
> (Ministry of the Interior, 2013: 3)

The connection between do-it-yourself democracy and larger trends is highly significant. Due to the economic crisis, many European countries are implementing austerity measures and cuts in public policy, alongside longer trends of welfare retrenchment. The idea is that citizens take responsibility and organise themselves to fill in gaps left by government spending cuts in healthcare, education, employment and neighbourhood governance (Newman and Tonkens, 2011;

Kleinhans *et al.*, 2013). This entails a clear appeal to citizens to come into action, because their efforts are assumed to result in alternatives that may be more (financially and administratively) sustainable than state-based services and facilities. Promoting responsibility and active citizenship have been among the core objectives of Dutch governments for years (Hurenkamp *et al.*, 2011). Crucially, however, the retrenchment of the Dutch welfare state, the financial crisis and the resulting austerity regime in the Netherlands and other EU countries have increased the sense of urgency that citizens should step up *together* to bridge the gaps created by government spending cuts. This is where the notion of sustainable communities enters the equation. Since 2008, the Netherlands has witnessed a surge of local initiatives, ranging from cooperatives in the field of energy, healthcare and food (Ministry of the Interior, 2013; Van der Schoor and Scholtens, 2015) to community-based social enterprises and other entrepreneurial forms of active citizenship (Ham and Meer, 2015; Kleinhans *et al.*, 2015). The final section of this chapter reviews the potential of community-based social enterprises for making local communities more sustainable.

Sustainable housing and communities indicators

Based on Tables 1.1 and 1.2 in Chapter 1, I will discuss a number of indicators for sustainable housing and communities in the Netherlands. First of all, the high population density is reflected in the distribution of population by dwelling type. Only 16 per cent of the population live in detached houses; less than half of the EU28 average. The proportion of people in flats is also relatively low (19 per cent). On the other hand, 60 per cent live in semi-detached houses, which is high compared to the EU28 average of 24 per cent. The prosperity of the Dutch is reflected in the distribution of population by housing tenure. Partly as a result of fiscal incentives for home ownership, the housing cost overburden rate for owners with a mortgage or loan is relatively high: 13 per cent versus an EU28 average of 8 per cent (see Table 1.2). By contrast, the figure for the rented sector is below average. Only 15 per cent are at risk of social exclusion, compared with 25 per cent in the EU28 (Eurostat, 2013). Generally, the quality of the Dutch housing stock is high, especially the quality of social rented housing, for which the country is renowned. Overcrowding is not an issue in the Netherlands, which has one of the lowest overcrowding rates in Europe at just 3 per cent.

On a country level, we can use the data provided by the extensive Monitor Sustainability Netherlands (MSN; *Monitor Duurzaam Nederland*), which is maintained by the government body Statistics Netherlands (CBS, 2014). This monitor uses the WCED definition of sustainability as explained at the start of this section. Based on this definition, MSN identifies three 'dashboards' that each visualise the most fundamental aspects of sustainable development:

1.   Quality of life in the present-day situation.
2.   The (use of) resources, which provides an impression of the extent to which future generations will be able to sustain a certain quality of life.

3.　The relative position of the Netherlands in the world, in terms of the impact of Dutch aims for prosperity and quality of life on the rest of the world.

For our perspective on sustainable housing and communities, the first aspect (quality of life) is the most relevant. This dimension consists of three main themes that are operationalised into further indicators:

1.　Welfare and material prosperity: compared to the base year 2003, the consumption expenditure index peaked in 2008 (index value 108) and decreased to 103 in 2013, due to the financial crisis. For the same reason, general satisfaction with life decreased in 2012. However, both index values are still relatively high when compared to other EU countries.
2.　Personal characteristics (health, housing, education, leisure, mobility, social security and pensions).
3.　Environmental factors (safety, social inequality, social participation and trust, institutions, nature and air quality).

(CBS, 2014: 20)

In terms of housing satisfaction, we see a positive trend between 2000 and 2014, and a relatively high position compared to other EU countries. Quality of housing has remained at roughly the same position in international perspective. Developments in mobility (home–work travel), financial security and structural unemployment have a negative impact on overall sustainability. For the latter two factors, the trends are directly related to the impacts of the GFC. As for mobility, Dutch citizens spend a lot of time commuting between home and work, compared residents of other European countries. Using several transport modes, Dutch employees travel on average 17.5 kilometres a day to and from their place of work. Approximately 75 per cent of this travel is done by car, 15 per cent by public transport (mostly train) and 6 per cent by bicycle. The average commuting time in 2015 was 26 minutes (CBS, 2015). The Netherlands stands out positively in terms of structural unemployment (despite the negative trend since 2000), probably because the GFC-related trends in unemployment and social security have been even more negative in other EU countries.

In terms of environmental trends, there are few changes in impact on sustainability, but at the same time the Netherlands has a relatively high international position. This applies to income inequality, social participation and trust, and trust in institutions (including voting rates during elections). There are a few remarkable exceptions to this rule, however. First, feelings of safety have decreased since 2000, but the relative international position of the Netherlands remains high in this respect. Crime victimisation has decreased, but this seems not to have affected the international position. In terms of managing and conserving nature reserves, the Netherlands' performance is relatively low compared to other EU countries (CBS, 2014). This is also reflected in the Netherlands' Environmental Protection Index score of 77.8 in 2014 (see Table 1.1), which is above the EU28 average but below the EPI scores of Germany, Norway, Spain, Sweden and Switzerland.

Barriers, drivers and challenges for sustainable communities in the Netherlands

As in many other countries, the GFC has reduced the prospects for achieving sustainable communities in the Netherlands. Several budget cuts and welfare state reforms are not only crisis-oriented, but also due to the government's determination to comply with the '3 per cent rule' of the European Commission. The Netherlands has complied with this rule since 2014, but this does not mean that the government has money to spend on creating or maintaining sustainable communities. A clear and relevant example is the fact that the Rutte I government (2010–2012) cut all funding for the Priority Neighbourhoods Approach, the backbone of Dutch neighbourhood regeneration policy. From a sustainability point of view, this programme was relatively successful in creating more variation in housing sizes, types, quality, tenure in target areas, improving public spaces, increasing housing career opportunities for middle-income households, and de-concentration of high shares of deprived residents (see Kleinhans, 2012). Its termination created substantial barriers with regard to the management and financing of area-based approaches towards achieving sustainable communities. In the same vein, housing associations played a fundamental role in terms of restructuring the social housing sector and increasing its energy efficiency, in particular through improving insulation, smart heating systems (e.g. heat pumps) and the use of sustainable energy through solar panels. Research has shown that some modest progress has been made by housing associations in this respect (Filippidou *et al.*, 2014). Further progress is significantly hindered by the lack of investment opportunities for housing associations.

Other barriers are related to legal changes. The functioning of housing associations has come under scrutiny as a result of allegations of illegal state support (Priemus and Gruis, 2011) and the disclosure of fraud and financial misconduct among large housing associations in the Netherlands. The most infamous example was Vestia, the largest housing association in the country, which came close to bankruptcy in 2012 as a result of investing in risky derivatives. Consequently, public opinion turned strongly against the housing associations, partly because their CEOs tend to receive huge salaries and drive expensive company cars (see Boelhouwer and Priemus (2014) for a detailed discussion). A Parliamentary Inquiry on Housing Associations documented all of the abuses in the social housing sector and issued several recommendations (Van Vliet *et al.*, 2014), some of which were included in the new Housing Act (*Woningwet*).

This act, established in July 2015, has significantly limited the tasks, competences and investment capacities of housing associations. First, the legal core business of housing associations is (again) limited to building and managing social rented housing for low-income households. Commercial project development is out of their scope. The only exceptions are efforts to regenerate deprived neighbourhoods and population shrinkage areas (*krimpgebieden*) by means of urban restructuring – that is, the demolition of social rented housing and construction of more expensive rented and owner-occupied housing. Housing associations may still try to improve the liveability (*leefbaarheid*) of neighbourhoods through social investment, but the scope will be strictly limited by performance agreements with local authorities. A second change is that housing associations must formally split their social and commercial activities in both legal and financial terms, which significantly increases their administrative burden. Finally, tenants in social housing will have much more formal influence, for example on performance agreements between housing associations, local authorities and tenants' organisations. The new Housing Act formally enables groups of tenants to establish housing cooperatives by taking over a (small) part of the stock of housing associations. However, this option is highly complex in practice, with the result that it has not yet been employed.

There are a number of other challenges for sustainable communities. As in the case of housing associations, local authorities are struggling with considerably reduced investment capacities while facing more responsibilities due to welfare state reforms and decentralisation policies. Not surprisingly, the call for active citizenship in various domains of public life has increased since the start of the economic crisis, particularly in terms of making citizens more self-reliant and less dependent on the welfare state. However, the path to active citizenship that contributes to sustainable communities is fraught with problems. A key challenge is that the required level of self-organisation is neither evenly distributed in space and time nor always democratically managed. 'Community-based self-organization is in fact an articulation of the diversity of issues, lifestyles, organizations and spatial interests within urbanized areas' (Boonstra and Boelens, 2011: 117). Many scholars have asserted that control of community assets, resources and networks but also the ability to self-organise is disproportionately more often available among privileged communities and community members (Kisby, 2010; Bailey, 2012; Uitermark, 2015). Self-organising practices of more resource-rich residents can exclude resource-poor residents, because the latter lack the necessary skills and the cultural and social capital. In other words, preventing social exclusion is one of the main challenges for sustainable communities based on the bottom-up self-organisation of citizens.

A final challenge for achieving sustainable communities in the Netherlands lies in their spatial demarcation. While everyone talks about the importance of sustainable communities, there is no clear idea of which residents are part of a certain community, and which residents are not. Policy-makers have long tried to increase participation levels in spatial planning policies, but have continually failed in this aim (Boonstra and Boelens, 2011). A key reason is that

> the whole notion of coherent, identifiable, spatially defined, communities is debatable; there is no guarantee that locality-based activities will 'bring together' residents; most people in deprived areas do not get involved locally, even within the context of well-funded area-based initiatives: they have other priorities.
>
> (Lawless, 2011: 59)

In sum, the notion of sustainable communities always begs the question: who is part of the community and who is not? I discuss this issue further in the next section.

The potential of community enterprises to create sustainable communities

This section will focus on community enterprises (CEs), as they are potentially powerful examples of bottom-up active citizenship that can contribute significantly to the concept of sustainable communities. CEs are community-based, owned and managed by residents, and develop non-profit activities to aid the regeneration of a particular neighbourhood or community (Bailey, 2012). They often arise through the perception of serious deficiencies in a particular area, such as deprivation, poor health, inadequate housing or a lack of community facilities, for which existing agencies are unlikely to provide solutions. These perceptions motivate individuals and groups to join forces and set up an organisation to generate solutions (Bailey, 2012: 26–27). Many CEs in the United Kingdom own assets such as land or buildings from which benefits arise (Aiken *et al.*, 2011).

In the Netherlands, the growing interest in British CEs sparked off a national programme of experiments with a Dutch equivalent – that is, *bewonersbedrijven*. I have led a three-year panel study that has monitored fourteen new *bewoners-bedrijven* (Kleinhans *et al.*, 2015). Below, I use the definition of 'sustainability' in the Bristol Accord (see above) to assess CEs' potential with regard to making communities (more) sustainable.

In terms of meeting the *needs of existing (and future) residents*, Dutch CEs conduct a wide range of activities, claiming that they address residents' needs for which existing agencies are unlikely to provide solutions. For instance, several CEs rent out rooms or office space in former primary school buildings, primarily to entrepreneurs, artists and students. Other CEs focus on taking over or competing for services and tasks that were previously conducted by local authorities or housing associations, such as managing and maintaining public spaces and parks, maintaining and cleaning staircases in social rented apartment buildings, and projects for reintegration of unemployed social service clients into paid employment. The government-driven philosophy of substitution (i.e. residents taking over) is one of the main reasons why we have found many CEs taking on services that were formerly provided by the public sector, such as libraries, food and clothing banks and shops that specialise in recycled goods, as well as maintaining playgrounds and even a pony stable. A third line of activities concerns small-scale services, such as a grocery service, cheap catering facilities, a low-cost taxi service, a bike-repair store, a sewing atelier and counselling services (Kleinhans *et al.*, 2015).

In terms of contributing to a *high quality of life* and *offering good service for all*, the questions pertain to who benefits from the CEs' activities and to what extent they offer 'good service'. In the first place, benefits accrue to renters of the cheap rooms and affordable office/working space, clearly meeting the needs of a group of low-income renters, such as students, entrepreneurs and people working in the creative sector, who often struggle to find affordable working spaces close to where they live. In the case of projects for reintegration of unemployed social service clients and trajectories for providing meaningful activities for (temporarily) disabled and unemployed residents, the benefits accrue to those clients themselves. CEs that offer cheap restaurants or hire workspaces to caterers provide very affordable ways of 'eating out' for low-income residents who would otherwise lack access to healthy meal opportunities at a reasonable distance. These restaurants function as social meeting places for neighbourhood residents. In the case of CEs maintaining public spaces, parks and buildings, the benefits accrue mostly to neighbourhood residents through positive effects on liveability and higher involvement, with volunteers' efforts acknowledged and appreciated by their fellow residents.

In terms of being *well planned and run*, several difficulties arise in the practice of CEs. First, they all struggle to create a sound business model, which is a *sine qua non* for creating a secure income stream so that social objectives can be achieved. We found that CEs whose business model is based on exploiting real estate assets are much better equipped to achieve this requirement than CEs whose business model is based solely on service provision (Kleinhans *et al.*, 2015). A second concern is that CEs are highly dependent on volunteers. As with many voluntary associations, volunteers are difficult to recruit and maintain. A specific problem in this context is that there are legal barriers for paying an allowance to unemployed volunteers who are on social benefits. Volunteers who accept financial allowances without consulting with the social security office risk losing their social benefit payments. In sum, serious questions can be raised with regard to the financial and staff sustainability of CEs.

In terms of *inclusiveness*, the picture is unclear. On the one hand, CEs appear to be inclusive for residents whose needs may have been harmed by current austerity measures. On the other hand, the notion of community implies both inclusion and exclusion of groups or individuals. By focusing on a specific target group, CEs may exclude other residents. Furthermore, the management of CEs may unintentionally foster exclusion because decisions have to be made with regard to the (financial) continuity of CEs. There is evidence that hierarchical decision-making processes may be appropriate to deliver services and to strengthen financial viability, but may simultaneously prevent excluded groups from participating (Teasdale, 2010). However, it is too early to say whether forms of social exclusion are a significant issue in the management and outcomes of Dutch CEs.

Conclusions

What are the current and future prospects for sustainable housing and communities in the Netherlands? Ultimately, the answer to this question is highly ambivalent. In terms of scores on a range of sustainability indicators, the position of the Netherlands compared to other EU countries is generally very good, especially in terms of housing quality, housing satisfaction, income, material prosperity, structural unemployment and the country's Environmental Protection Index score. In parallel with many other EU member states, the Netherlands has worked hard to implement EU directives on various sustainability measures that had to be operationalised at member-state level. For example, the Agenda Sustainability (*Agenda Duurzaamheid*) and the Energy Agreement for Sustainable Growth (*Nationaal Energie Akkoord*) reveal the commitment of various stakeholders to achieve sustainability. The effectiveness of these arrangements is still under discussion. Perhaps the most succinct conclusion on this matter has been drawn by Statistics Netherlands, the government body that is responsible for Monitor Sustainability Netherlands (*Monitor Duurzaam Nederland*), which states that 'the quality of life in the Netherlands is high, but the way in which this quality is achieved does take a relatively large claim on natural resources and, to a lesser extent, also on human capital' (CBS, 2014: 7).

The other side of the coin relates to larger societal trends and the ability of Dutch key actors to work on sustainable housing and communities in the Netherlands. There are several macro tends that are extremely difficult to address: the ageing of the population; the associated rising costs of pensions, healthcare and population shrinkage; and problems with housing affordability, especially for young people. Together, these forces pose substantial challenges to maintaining facilities and seriously complicate attempts to achieve sustainable housing and communities. Local authorities and housing associations used to be the key actors in policies contributing to sustainable housing and communities, such as urban regeneration and energy efficiency. Due to budget cuts and legal changes, their investment capacity has significantly decreased since 2008. The Dutch government perceives active citizenship as a viable alternative to state-based welfare provision, because the latter is no longer sustainable for financial and other reasons. This tendency has been criticised by pointing at the problematic notion of 'community', which entails both social inclusion and exclusion, and the paradox of self-organisation – that is, the fact that the ability to self-organise is disproportionately more often available in more privileged communities and among more affluent community members.

In the final part of this chapter, I sought to analyse the extent to which community enterprise has the potential to contribute to sustainable communities. Research evidence shows that this potential is very modest, at least in the short term. While we found some evidence that CEs can contribute to sustainable communities in terms of quality of life and offering good services, serious questions have to be raised with regard to the financial sustainability and staffing capacity of CEs (Kleinhans *et al.*, 2015). The history of CEs in Britain has shown that it takes many years – if not decades – to establish more or less stable and successful community enterprises (Bailey, 2012). In other words, the current modest potential of Dutch CEs may increase over time. Future research is needed to establish whether that potential will be realised.

References

ABF Research (2012) *Systeem Woningvoorraad 2012 – SYSWOV* [*System Housing Stock 2012*]. Delft: ABF Research.

Aiken, M., Cairns, B., Taylor, M. and Moran, R. (2011) *Community Organisations Controlling Assets: A Better Understanding*. York: Joseph Rowntree Foundation.

Bailey, N. (2012) 'The role, organisation, and contribution of community enterprise to urban regeneration policy in the UK', *Progress in Planning*, 77: 1–35.

Bergeijk, E. van, Kokx, A., Bolt, G. and Kempen, R. van (2008) *Helpt Herstructurering? Effecten van Stedelijke Herstructurering op Wijken en Bewoners* [*Does Restructuring Help? Outcomes of Urban Restructuring on Neighbourhoods and Residents*]. Delft: Eburon.

Boelhouwer, P. (2014) 'Market orientated housing policies and the position of middle-income groups in the Netherlands', *Journal of Housing and the Built Environment*, 29: 729–739.

Boelhouwer, P. and Priemus, H. (2014) 'Demise of the Dutch social housing tradition: impact of budget cuts and political changes', *Journal of Housing and the Built Environment*, 29: 221–235.

Boonstra, B. and L. Boelens (2011) 'Self-organization in urban development: towards a new perspective on spatial planning', *Urban Research and Practice*, 4(2): 99–122.

Brenner, N. (2004) *New State Spaces: Urban Governance and the Rescaling of Statehood*. Oxford: Oxford University Press.

CBS [Statistics Netherlands] (2014) *Monitor Duurzaam Nederland 2014 – Indicatorenrapport* [*Monitor Sustainable Netherlands 2014 – Indicator Report*]. The Hague: Statistics Netherlands.

CBS [Statistics Netherlands] (2015) *Transport and Mobility 2015*. The Hague: Statistics Netherlands.

Dijkhoff, T. (2014) 'The Dutch Social Support Act in the shadow of the decentralization dream', *Journal of Social Welfare and Family Law*, 36(3): 76–294.

Eurostat (2013) 'At risk of poverty or social exclusion in the EU28'. Eurostat news release, STAT/13/184, 5 December 2013.

Filippidou, F., Nieboer, N. and Visscher, H. (2014) 'The pace of energy improvement in the Dutch non-profit housing'. Paper presented at World SB14, Barcelona, 28–30 October. Available at www.irbnet.de/daten/iconda/CIB_DC28178.pdf, accessed 11 July 2016.

Gavrilow, L. and Heuveline, P. (2003) *Aging of Population: The Encyclopaedia of Population*. New York: Macmillan.

Gruis, V., Visscher, H. and Kleinhans, R. (eds) (2006) *Sustainable Neighbourhood Transformation: Perspectives on Housing Demolition and Development*. Amsterdam: IOS Press.

Ham, M. van and Meer, J. van der (2015) *De Ondernemende Burger. De Woelige Wereld van Lokale Initiatieven* [*The Entrepreurial Citizen: The Turbulent World of Local Initiatives*]. Utrecht: Movisie.

Hoekstra, J. (2003) 'Housing and the welfare state in the Netherlands: an application of Esping-Andersen's typology', *Housing, Theory and Society*, 20(2): 58–71.

House of Representatives (2013) *Memorie van Toelichting Bij de Regels Over Het Beschermen en Benutten van de Fysieke Leefomgeving – Omgevingswet*. Year 2013–2014, File 33 962, No. 3. The Hague: House of Representatives.

Hurenkamp, M., Tonkens, E. and Duyvendak, J. W. (2011) 'Citizenship in the Netherlands: locally produced, nationally contested', *Citizenship Studies*, 15(2): 205–225.

Jessop, B. (2002) *The Future of the Capitalist State*. Cambridge: Polity Press.

Kisby, B. (2010) 'The Big Society: power to the people?', *Political Quarterly*, 81(4): 484–491.

Kleinhans, R. (2012) 'A glass half empty or half full? On the perceived gap between urban geography research and Dutch urban restructuring policy', *International Journal of Housing Policy*, 12: 299–314.

Kleinhans, R., Ham, M. van and Doff, W. (2013) 'Transferring British community enterprises to the Dutch context: the Big Society in disguise or a promising concept for residents' self-organisation?' Paper presented at the 25th ENHR Conference: Overcoming the Crisis, Integrating the Urban Environment, Tarragona, 25–28 June.

Kleinhans, R., Doff, W., Romein, A. and Ham, M. van (2015) *Project Kennisontwikkeling Experiment Bewonersbedrijven - Eindrapport* [*Project Experiment and Knowledge Development on Community Enterprises – Final Report*]. Delft: Faculty of Architecture and the Built Environment, Delft University of Technology.

Lawless, P. (2011) 'Big Society and community: lessons from the 1998–2011 New Deal for Communities Programme in England', *People, Place and Policy Online*, 5(2): 55–64.

Ministry of Infrastructure and the Environment (2011) *Agenda Duurzaamheid; een Groene Groeistrategie voor Nederland* [*Agenda Sustainability: A Green Growth Strategy for the Netherlands*]. The Hague: Ministry of Infrastructure and the Environment.

Ministry of Infrastructure and the Environment (2013) *De Doe-Democratie. Kabinetsnota ter Stimulering van een Vitale Samenleving* [*The DIY Democracy: White Paper for Stimulating a Vital Society*]. The Hague: Ministry of Infrastructure and the Environment.

Ministry of the Interior (2013) *Burgercoöperaties in Opkomst. De Veranderende Relatie tussen Samenleving en Overhead* [*Citizen Cooperatives on the Rise: About the Changing Relations between Society and the Government*]. The Hague: The Countryside Network and the Ministry of the Interior.

Musterd, S. (2014) 'Public housing for whom? Experiences in an era of mature neo-liberalism: the Netherlands and Amsterdam', *Housing Studies*, 29(4): 467–484.

Office of the Deputy Prime Minister (ODPM) (2005) *Sustainable Communities: People, Places, and Prosperity* [Bristol Accord]. London: ODPM.

Priemus, H. (2003) 'Dutch housing associations: current developments and debates', *Housing Studies*, 18(3): 327–351.

Priemus, H. and Gruis, V. (2011) 'Social housing and illegal state aid: the agreement between European Commission and Dutch government', *European Journal of Housing Policy*, 11(1): 89–104.

Social and Economic Council (SER) (2013) *Summary of: Energy Agreement for Sustainable Growth (Energieakkoord voor duurzame groei, 06-09-2013)*. The Hague: SER.

Teasdale, S. (2010) 'How can social enterprise address disadvantage? Evidence from an inner city community', *Journal of Non-profit and Public Sector Marketing*, 22(2): 89–107.

Tinker, A., Ginn, J. and Ribe, E. (2013) *Assisted Living Platform: The Long Term Care Revolution: A Study of Innovatory Models to Support Older People with Disabilities in the Netherlands*. Case Study 37. London: Housing Learning and Improvement Network.

Uitermark, J. (2015) 'Longing for Wikitopia: the study and politics of self-organisation', *Urban Studies*, 52: 2301–2312.

Van Daalen, G. and Van der Land, M. (2008) 'Next steps in choice-based letting in the Dutch social housing sector', *International Journal of Housing Policy*, 8: 317–328.

Van der Schoor, T. and Scholtens, B. (2015) 'Power to the people: local community initiatives and the transition to sustainable energy', *Renewable and Sustainable Energy Reviews*, 43: 666–675.

Van Vliet, R., Oskam, P., Hachchi, W., Mulder, A., Groot, E. and Bashir, F. (2014) *Parlementaire Enquête Woningcorporaties* [*Parliamentary Enquiry on Housing Associations*]. The Hague: House of Representatives.

Varady, D., Kleinhans, R. and Ham, M. van (2015) 'The potential of community entrepreneurship for neighbourhood revitalization in the United Kingdom and the United States', *Journal of Enterprising Communities: People and Places in the Global Economy*, 9(3): 253–276.

Verhoeven, I. and Tonkens, E. (2013) 'Talking active citizenship: framing welfare state reform in England and the Netherlands', *Social Policy and Society*, 12: 415–426.

Wiles, J., Leibing, A., Guberman, N., Reeve, J. and Allen, R. (2012) 'The meaning of "aging in place" to older people', *Gerontologist*, 52(3): 357–366.

World Commission on Environment and Development (WCED) (1987) *Our Common Future*. Oxford: Oxford University Press.

World Health Organisation (WHO) (2007) *Global Age-Friendly Cities: A Guide*. Geneva: World Health Organisation.

8   The United Kingdom

Glen Bramley

Introduction

The United Kingdom (UK) is one of the larger European countries and economies, operating outside the Euro Zone, which enjoys an economic living standard and growth rate comparable with the other major European countries. In 2014 its per capita GDP (on PPP basis) was 28 per cent below the USA and 14 per cent below Germany, but only 2 per cent below France, 6 per cent above Japan and 11 per cent above Italy (IMF, 2015). Growth rates over the period since 2003 compare favourably with European peers, with an average of 1.6 per cent against 0.8 per cent for the Euro Zone, 1.1 per cent for Germany and 1.0 per cent for France, but below the USA's 1.9 per cent. Growth was relatively high in the early 2000s, but dipped sharply in the 2008–2009 crisis period and was slow to recover, although recovery quickened in 2013–2014 (Bozio *et al.*, 2015).

Although government ministers boast of a relatively good economic performance since 2013, there are in reality several factors which raise doubts about the sustainability of this, including a relatively large and persistent government deficit and an equally large balance of payments current account deficit, both around 5 per cent of GDP, high government and personal debt, and flatlining productivity and earnings since 2008. The exceptionally high population growth rate, one of the highest in Europe (ONS, 2014a), flatters the GDP growth and helps to explain why it is not translating into living standards.

The UK has historically operated as a unitary state, but devolution to the countries of Scotland, Wales and Northern Ireland has advanced since 1997 and, following the relatively close referendum result on Scottish independence in 2014, will advance further in the coming period. Most matters relating to housing and urban policy have been substantially devolved since 1999, with further steps to devolution in terms of fiscal arrangements and some welfare benefits and regulatory regimes in train. Because devolved policies have progressively diverged, not least in housing-related areas like homelessness (Pawson and Davidson, 2008), it becomes more difficult to present a succinct single description of the UK's experience. Therefore, this chapter focuses particularly on England, where housing pressures are greatest.

The UK's welfare regime is conventionally labelled as 'liberal', with a dualistic housing rental housing system, but this perhaps glosses over a historical legacy that was more social democratic in character. It was principally after the Thatcher government era of the 1980s that the neoliberal ideology gained fuller force and housing was reshaped in a predominantly marketised form, in parallel with harsher regimes in terms of social security and employment.

The neoliberal tendencies since 1980 have been reflected in a marked shift towards greater inequalities of income and wealth, such that the UK is now one of the most unequal advanced countries in Europe, although this shift has also been driven by structural economic changes associated with the move to a post-industrial service economy, reflected in pre-tax income inequality. In 2010, the UK had the fifth most unequal incomes of thirty-one industrialised countries, behind only Chile, the USA, Israel and Portugal (based on Gini coefficients for disposable income; Hills, 2015: Figure 2.6). Wealth inequality is also relatively high, with the top decile owning 70 per cent in 2010, compared with 80–90 per cent in the gilded age of 1870–1910 – comparable with 63 per cent European average, 59 per cent in Sweden, and 75 per cent in the USA (Picketty, 2014: ch. 10). There is a generational aspect to this as well, with the older/retired generation tending to see reducing poverty incidence and increasing wealth while younger, working-age households see more poverty and financial stress, including housing affordability problems. The latter are also the principal targets of welfare cuts.

Britain is a mature post-industrialised, long-urbanised society which gained a legacy of reasonable housing (mainly suburban 'houses' rather than apartments) from the inter-war and post-war housing booms and the legacy of large-scale public house-building up to the 1970s. Having lived off this legacy for twenty-five years when demographic demands were moderate (Bramley, 2007), England (in particular) has experienced an unprecedented increase in population through immigration, especially over the last fifteen years. This population growth is radically higher than that seen over the previous forty years (i.e. since the 1960s). The UK's population grew by 5 million between 2001 and 2013, *half of the total growth of 10 million since 1964*. The annual growth rates of 0.7–0.85 per cent since 2004 compare with 0.4–0.5 per cent in the late 1960s, around zero in the 1970s, 0–0.2 per cent in the 1980s, and 0.2–0.5 per cent in the 1990s. England generally has a higher growth rate than the UK as a whole (0.79 per cent versus 0.73 per cent over the last decade). Remarkably, the UK had higher absolute population growth than any other EU state in 2012–2013, and a higher percentage increase than any but Sweden (ONS, 2014a). This growth has been driven predominantly by migration, with net international migration accounting for more than half in all years from 1999 to 2011. The accession of eight new EU member states in 2004, with immediate free access to the UK's labour market, caused an upward hike in numbers, but this has not been the only factor. There were earlier increases in the mid- to late 1990s, and various streams of migration have all contributed to this outcome.

Unfortunately, Britain (and especially England) has a chronic undersupply of new house-building, over an extended period, compounded by a generally unresponsive ('inelastic') supply system in the face of surging demand (Barker, 2004; Bramley, 2007). Demand rose rapidly in the early 2000s, with the fundamentals of strong income growth, massive demographic growth, low/stable interest rates combined with liberal credit terms, contributing to boom conditions. Affordability ratios remained high through the recession and during the aftermath, with the UK fourth out of twenty-nine countries in 2014 (IMF, 2015). Britain is somewhere in the middle of the pack, in terms of OECD countries, in having seen a 15 per cent real house price fall since 2007. Generally more financially deregulated systems have experienced more market volatility through the crisis (Whitehead *et al.*, 2014: Table 1). Britain saw a substantial fall in housing output in this period (–42 per cent, 2007–2012), in contrast to some of the more stable countries (Germany –3 per cent; Sweden –16 per cent), but less spectacularly than Spain (–81 per cent), Ireland (–88 per cent) or USA (–68 per cent). Perhaps more striking was the finding that Britain had one of the lowest *levels* of housing completions per adult population of all countries in that comparison, even in the peak year of 2007 (Whitehead *et al.*, 2014: Table 5).

The causes of chronic undersupply include the operation of the land use planning system (Barker, 2004; Bramley, 2007, 2012a) and also (arguably) the structure and behaviour of the house-building industry (Leishman, 2015). Planning is mainly based at the local level, and, reflecting public sentiment, tends to be particularly restrictive in the high-demand south of England (Bramley, 2012a, 2013). Embedded within the planning system since the 1990s has also been a strong ethos of 'sustainable development' (DCLG, 2012). This may be an indirect route to unlocking the supply conundrum – if local communities were convinced they could ensure that new communities would be sustainable, they would be more likely to allow them to be built.

The housing system in the UK

At national government level primary responsibility for housing policy rests with the Department of Communities and Local Government (DCLG) in England, while it is devolved in the other UK countries. This department also holds responsibility for planning and urban regeneration, but not for other housing-related policies, including Housing Benefit/Allowances, energy efficiency and financial regulation (most of which are not devolved). Commentators generally agree that housing has not been a top political issue over recent decades, although since the mid-2000s it has secured more political and policy attention (Bramley 1997, 2007), and it did feature in the 2015 general election campaign in terms of offering to help 'generation rent' (see below) into home ownership. However, since 2010 the national government has adopted austerity fiscal policies, and in this context it seems mainly interested in housing policies which do not demand a lot of public expenditure or borrowing.

Local government was very important in the development and implementation of housing policy in the twentieth century, particularly between 1945 and 1975, when local authorities typically built half of all new housing and came to own about a third of the stock (over half in Scotland). Since 1975, this picture has been transformed by a range of policies and changes, including most famously the 'right to buy' council housing, but also spending limits, privatisation and reduced involvement in urban regeneration. Local authorities are still responsible for planning and housing in their localities, but they have limited resources for implementation. Also, in many cases since 1989, the main social housing stock has transferred to housing associations (regulated non-profit social housing providers), with most of the modest amount of new social housing built by these bodies since the 1980s. On the whole, these bodies seem to be successful in raising the quality of housing management and development, although there are lingering concerns about accountability and value for money (Pawson *et al.*, 2010).

Housing tenure in the UK is not very different from the EU average, with 65 per cent owner occupied compared with 67 per cent in the EU18 or 70 per cent in the EU28. The main difference is that social housing has a bigger share in the UK, as a legacy of the earlier post-war period, still accounting for 18 per cent compared with 11 per cent across the EU (both definitions). The private rented sector accounts for a correspondingly smaller share at 17 per cent, compared with 22 per cent in EU18 or 19 per cent in EU28.

Trends over time differ somewhat, however. In the UK, owner occupation has been falling since the early 2000s, and is now about 5 per cent below its peak, whereas across the EU it has been increasing. Social renting is in gentle decline, with 'right to buy' sales at a relatively low ebb in recent years, but set to increase again. Private renting has been on a strongly upward trajectory since the early 2000s, up about 5 per cent since 2005 and more or less doubling its share since the early 1990s, which contrasts with a slight decline in the EU. The growth of private renting is the biggest story, driven on the demand side by unaffordable owner occupation and demographics, and on the supply side by the popularity and financial ease of 'buy to let' investments among more affluent, older households.

New building for social renting has averaged about 25,000 per year in England and 5,000 per year in Scotland in recent years. The primary mechanism for new social building has been through housing associations, who receive capital grants from government to cover part of the cost and finance the remainder from private market borrowing and internal resources. Grant availability has declined since 2010, leading to higher rents and, prospectively, a sharp reduction in new social house-building in England, with more emphasis on assisted home ownership. Scotland has not followed this route. An inclusionary planning mechanism (known as 'Section 106') has been used to support a majority of affordable housing supply in England over the last decade, bringing subsidy from lowered land values while also supporting 'mixed communities'.

Social housing in the UK has effectively been a secure tenancy, with allocation according to need, and low-income households have traditionally seen this as the long-term tenure of choice if they can get into it. By contrast, private renting has been deregulated since 1989 and the normal form of tenancy is a six-month shorthold, with no real long-term security. In the early 1990s, private renting was more of a niche tenure for young mobile households, and this was less of an issue than it is now, with a much wider range of households obliged to seek accommodation within it, often for long periods. At the time of writing the post-2015 Conservative government was legislating to make social tenancies fixed term and to raise rents to market levels for higher-income tenants. While this would appear to be taking social housing in England closer to the 'liberal' regimes of Australia and the US, the government has encountered difficulties in applying these policies to housing associations. Again, these policies have not been adopted in Scotland.

Another distinct feature of the British system is the homelessness legislation, in operation since 1977. This has essentially operated to give a right of access to social housing for households who are unintentionally homeless or at risk of losing their home if they are in priority groups, including families with children and vulnerable adults. (In Scotland, this has been extended to all homeless/at risk households.) As a consequence, a large proportion of social housing lettings go to homeless households, most of whom are unemployed or on very low incomes, and often have additional social problems and vulnerabilities. In England, especially, there has been a drive to reduce the numbers of rehoused homeless, through rigorous application of prevention measures, diverting some of these households into private renting or in some cases seeking to reconcile families to enable people to continue to live with their parents or relatives (Fitzpatrick *et al.*, 2015). Scotland has followed a different path, emphasising a more generalised right to rehousing and, where necessary, support, perhaps enabled by the easier balance of supply and demand. Consequently, this is another area where devolution has led to progressive divergence within the UK.

The other really distinctive feature of the UK housing system is the strong role played by the Housing Benefit (HB) system since the 1970s. This is a means-tested individual subsidy for renters which meets the total rent for most social renters with very low incomes, with a tapering off of subsidy as incomes rise. For private renters, since 2006 the system now known as Local Housing Allowance (LHA) has been modified to be based on local area guideline rents (set initially at the 30th percentile of market rents across broad market areas) for the relevant size of accommodation, rather than actual rents, with further limits on younger singles based on the rent of a room in shared accommodation. The HB/LHA system reflects a long-standing tradition in the UK's social security system to separate the housing element of basic income support from general income maintenance, and to reflect actual costs (which vary greatly between individuals and geographical areas).

This system has many consequences, the significance of which is heightened by the extent of income inequalities and low-income poverty alluded to above. Low-income renters expect to have their housing costs met, but they are kept close to a low poverty line with a high rate of withdrawal of benefit if they manage to increase their earnings. This is alleged to reinforce higher levels of worklessness, especially among social tenants. Social housing providers expect their rental incomes to be easily collectable (indeed, they are often paid direct), and this also reassures their private lenders. However, the costs of HB/LHA have tended to escalate over time, particularly in recessions and with rising rents, and reflecting widening inequalities. The cost of this is fiscally unsustainable, so there is strong pressure for reform, seen particularly in changes to LHA since 2011 and more recently in changes in the social sector, particularly the 'bedroom tax', which reduced benefit for social tenants with spare bedrooms. The latter proved highly unpopular and politically controversial (Gibb, 2014). Further reforms propose to merge HB/LHA with other benefits for working-age households in a new system known as 'Universal Credit'.

Turning back to owner occupation, although the UK market has revived following the prolonged period of crisis (2008–2012), there are concerns that access to this tenure will remain more difficult in future for those who do not have access to significant family wealth. In the mid-2000s, regulation of mortgage lending was 'light touch' and unsustainable lending practices developed, such as heavy use of interest-only loans and 'self-certified' incomes. The crisis itself was characterised by much tighter lending criteria, and in particular the imposition of large deposit downpayment requirements (e.g. 20 per cent deposits, contrasted with the UK norm of 5 per cent post-1980s). Since the crisis, regulation has been strengthened and income-affordability tests in particular are tougher. In consequence, many middle-income working households continue to be unable to afford to buy, particularly in the high-priced markets of London and the south of England (see Figure 8.1), giving rise to the phenomenon of 'generation rent'. The government has responded with various initiatives, including a 'help-to-buy' scheme targeted at new builds (so also promoting supply), utilising a shared equity model, and more recently a subsidised savings-bonus scheme and a crude shallow discount model called 'Starter Homes'. Critics have voiced concerns that these mechanisms simply serve to stoke up demand without necessarily increasing supply.

Sustainable communities and housing

Sustainable development has been a central theme of planning and urban policy since the late 1980s, and this has been paralleled by academic research seeking to clarify the concept of sustainability, develop indicators of sustainability outcomes and understand their relationships with patterns of urban development. Starting from the Brundtland definition – 'meeting today's needs without compromising future generations' ability to meet their needs' (WCED, 1987: 8) – there is general agreement that this comprises three strands: economic, environmental and social. Given our focus here on 'sustainable communities', it is the social strand which takes centre stage, but in practice it is always necessary to balance this with measures to secure and advance the environmental and economic dimensions as well.

In both UK and international academic literature there has been a proliferation of work exploring the concept and the reality of social sustainability, as well as its relationship with urban development. While initially it might be argued that the social dimension was less clearly defined, this work has served to put more flesh on the bones of the concept. In a number of contributions since 2009 this author has, with others, developed the argument that social sustainability comprises two main sub-dimensions, concerned on the one hand with 'social equity' and on the other with 'sustainability of community' (Bramley and Power, 2009; Dempsey *et al.*, 2009, 2012). For some, social sustainability has mainly been about equity, partly picking up on the 'intergenerational equity' perspective implicit in Brundtland and partly reflecting on the gross global disparities in human living conditions. Equity must be considered to render environmentalism a sustainable political strategy. Within the context of particular countries, such as those of Western Europe, social equity relates to urban form, mainly through issues of access to services, jobs and opportunities (including housing), particularly for the poorer sectors of society. However, we argued that social sustainability should also embrace broader questions about the functioning of society in an urban context, which one might characterise as the sustainability of community itself. This discussion has overlapped heavily with ongoing debates, research and policy concerns about social exclusion, social cohesion and social capital (Forrest and Kearns, 2001; Pierson, 2002; Coleman, 1988; Putnam, 2000; Mitlin and Satterthwaite, 1996; Fukuyama, 2000). This leads to a focus on a number of sub-domains, including: social interaction and networks; participation in collective group activities and structures; community stability versus mobility; pride in locality and sense of place; and safety and security (Wilson and Taub, 2006; Nash and Christie, 2003; Talen, 1999). The social sustainability perspective has also come to embrace or overlap with concerns about the effects of urban settings on key dimensions of human 'capabilities', notably health, and thence on broader outcomes of 'wellbeing' or 'happiness' (Wilkinson and Marmot, 2003; Barton *et al.*, 2003; Layard, 2005). Through the lens of health and wellbeing one can see both the equity/access and the social functioning of community contributing, but also economic functioning (avoiding material poverty) and key environmental issues, such as the health benefits of green space and from exercise through active travel.

Policy frameworks in the UK have increasingly taken these perspectives on board, notably in planning policy and practice, but also in allied fields including housing, local government, public health and transport. The UK Sustainable Development Strategy (DEFRA, 2005), which built on earlier initiatives (DETR 1997, 2001), provided an overarching framework with five guiding principles:

- living within environmental limits;
- a strong, healthy and just society;
- a sustainable economy;
- good governance; and
- sound, responsible science.

Within this, 'sustainable communities' were characterised as:

- active, inclusive and safe;
- well run;
- environmentally sensitive;

- well designed and built;
- well connected;
- thriving;
- well served; and
- fair for everyone.

This strategy is formally still in place, and the coalition government of 2010–2015 produced annual reports on its progress, which tended to focus narrowly on central department actions.

There is also a set of sustainability indicators maintained by the Office for National Statistics (ONS, 2014), which provides regular monitoring of progress over the medium and longer term across a wide range of indicators. These indicators are broadly grouped into economic, social and environmental strands. The social strand comprises indicators of healthy life expectancy (favourable trends), social capital (civic participation, social participation, social networks and trust – one of which is favourable short term), social mobility in adulthood (favourable) and housing provision (adverse). This strand of work also issues periodic reports with a more specific focus, for example Randall (2012) on 'where we live', which focused on satisfaction with living accommodation, importance of services and amenities, tenure, housing stock, housing conditions and the housing market, satisfaction with local area, access to the local environment and access to local services. Development of longer time series for some of these measures has been hampered by cutbacks in some surveys. While these data often refer to England, the Scottish government has also developed a quite elaborate 'outcomes' framework for policy, under the 'Scotland Performs' banner. Among the fifty national indicators are a number of relevance to this chapter, including:

- reduce traffic congestion;
- improve levels of educational attainment;
- increase physical activity;
- improve general health and mental wellbeing;
- reduce the proportion of individuals living in poverty;
- improve access to suitable housing options for those in housing need;
- increase the number of new homes;
- improve people's perceptions of their neighbourhood;
- increase cultural engagement;
- reduce Scotland's carbon footprint; and
- increase the proportion of journeys to work made by public or active transport.

While there is clearly a strong focus on aspects of sustainability in a range of government programme areas, such as health and transport, it is perhaps in the National Planning Policy Framework (NPPF; DCLG, 2012) that there is the clearest focus on relevant aspects (see Box 8.1).

Box 8.1   National Planning Policy Framework focus on sustainability

The purpose of the planning system is to contribute to the achievement of sustainable development. The policies in paragraphs 18 to 219, taken as a whole, constitute the Government's view of what sustainable development in England means in practice for the planning system.

There are three dimensions to sustainable development: economic, social and environmental. These dimensions give rise to the need for the planning system to perform a number of roles:

- an economic role – contributing to building a strong, responsive and competitive economy, by ensuring that sufficient land of the right type is available in the right places and at the right time to support growth and innovation; and by identifying and coordinating development requirements, including the provision of infrastructure;
- a social role – supporting strong, vibrant and healthy communities, by providing a supply of housing required to meet the needs of present and future generations; and by creating a high quality built environment, with accessible local services that reflect the community's needs and support its health, social and cultural well-being; and
- an environmental role – contributing to protecting and enhancing our natural, built and historic environment; and, as part of this, helping to improve biodiversity, use natural resources prudently, minimise waste and pollution, and mitigate and adapt to climate change including moving to a low carbon economy.

(DCLG, 2012: 2–3)

Under Sub-section 6, the NPPF emphasises the need to achieve a wide choice of high-quality homes, to ensure deliverable land supply to meet need and demand in full, to provide for a demographic mix of population as well as affordable housing, provision for 'self-building', reuse of empty buildings, including change of use, and the possibility of new settlements/urban extensions that follow Garden City principles. Sub-section 7 concerns 'good design', and emphasises 'sense of place', respect for local character and history, mixed uses, and attractive, comfortable and safe environments that are visually attractive. Sub-section 8 deals with promoting healthy, inclusive communities, and emphasises opportunities for meeting, including through mixed uses, strong neighbourhood centres and active street frontages, safe, accessible environments with clear, legible pedestrian routes and high-quality public space which encourages continual use. There is also an emphasis on local facilities and services, both planning for new provision and retention of existing, with a particular stress on adequate school places (currently a very fraught issue), and on public and recreational space.

Neighbourhood-level planning is particularly encouraged in this context. Sub-section 9 focuses on the protection of green belt land, an emphasis which arguably conflicts with the goals of meeting housing needs in the most sustainable fashion (Bramley and Watkins, 2014). Sub-section 10 focuses on meeting the challenges of climate change, flooding and coastal change, and urges local authorities to plan for new developments in locations and ways that will reduce greenhouse gas emissions. Scottish planning policy follows broadly similar lines.

Barriers, drivers and challenges for sustainable communities in the UK

The major theme developed here is that there is a serious imbalance between housing demand and supply in the UK, especially England. This section will use national survey sources, and some international comparisons, to draw out some aspects of this key issue.

Aggregate supply and demand

The first issue is that of the balance between overall demand and supply, and here the picture presented is very stark. The UK, and specifically England, faces almost the highest rate of increase in demand across Europe yet demonstrates one of the lowest rates of new build supply (Whitehead *et al.*, 2014). As outlined earlier, this buoyant demand is driven by migration and population growth, by recovering economic growth and by low interest rates. The pressure of demand is greatest in London, further boosted by international investor activity and financial sector bonuses, but is also high across the rest of the south of England. This imbalance has many consequences, picked up in more detail below, for affordability, the exacerbation of poverty, undesirable tenure shifts, housing needs and quality, and balanced communities. Intense pressure in London, for example, is fuelling rampant gentrification and, alongside welfare cuts, is increasingly displacing the poor from the central city (*Independent*, 2015). It is increasingly difficult to apply inclusionary policies for mixed communities when government, panicking about housing supply, allows established planning obligation policies to be watered down or overridden; or when political opportunism leads to the mandating of local authorities to sell their social housing in high-value areas to fund discounts to housing association tenants newly given the 'right to buy'.

It is conventional to assess housing shortfall by comparing actual and projected numbers of households and dwellings, but this naïve approach misses the point in a case like England. Official household projections show household growth rates falling (e.g. from 230,000 per annum in 2008-based projection to 206,000 per annum in 2012-based), but that is simply because limited housing supply is throttling household formation. In the 1970s and 1980s a large share of household growth was due to household formation (more adults living independently in smaller households). In the projected future for England 93 per cent of household growth is due to population increase, implying that net household formation has completely stalled. Young adult headship rates are falling and numbers of concealed households are rising.

Affordability and poverty

The picture on affordability is mixed. On the one hand, the European comparative indicators ('housing cost overburden rate') presented in Table 1.2 suggest that the UK has a peculiarly favourable position, second only to Ireland. Also, there is no doubt that, thanks to post-crisis low interest rates, many home owners have been protected from difficulties. On the other hand, different indicators give a somewhat different picture. For example, Bramley and Besemer (2013: Table 10) found the UK second or third worst within EU15 countries on four out of five affordability indicators tested on EU-SILC in 2009, including 'gross outgoings/gross income' and 'in poverty after but not before housing costs'. This shows that much depends on the particular indicator used, and the particular features of the British HB system mean that the 'overburden' indicator based on net costs and income is less revealing in this case (see also Bramley, 2012a). Affordability varies markedly between regions, tenure and age groups.

Figure 8.1 presents a conventional indicator relevant to access to home ownership and the general housing market – the house price: earnings ratio (at lower quartiles) by selected region. This shows the big upward hike from 2001 to 2007, the uneven effects of the crisis and recession, and the beginnings of recovery. At the end of the period, ratios are far above their 1990s levels in all regions, while ratios in London and the rest of the south are far above the northern and midland regions. It is true that low interest rates make such ratios more affordable for existing owners, but for potential new buyers they are inaccessible, given the tighter mortgage regulations that insist on significant downpayments and an affordability test based on twenty-five-year repayment mortgage with capability of withstanding a 2 per cent increase in interest rates.

We can use the Understanding Society Survey (USS) for 2009–2011 to derive an indicator based on a combination of objective ratios (gross housing rent/ mortgage cost to gross household income ratio over 25 per cent, *or* residual net income after housing costs less than 120 per cent of UK social security scale rate for household type) *and* subjective financial difficulty (problems paying for housing in the previous year). Affordability problems defined in this way affected 2.4 per cent of all households but this rose to 3.1 per cent in London, 3.9 per cent for private renters, 6.4 per cent for social renters, 4.0 per cent for under-forties (compared with 0.5 per cent for over-sixty-fives) and 13.6 per cent for lone-parent families. The high score for social renters is noteworthy, underlining that this tenure, in combination with HB, does not always solve affordability problems, particularly in the face of extensive and significant poverty.

Adverse tenure trends

Governments from across the political spectrum have generally favoured owner occupation and sought to promote it – the concept of 'tenure neutrality' has never

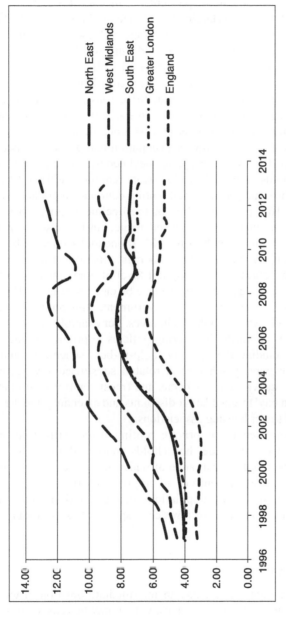

Figure 8.1 House price: earnings ratios by selected region, 1997–2013 (ratio of lower quartile price to lower quartile full-time earnings).

Source: DCLG, n.d.: Table 576.

had much political traction in the UK. In the academic and policy communities there have been ongoing debates about whether this normative preference is justified. Some arguments (particularly relevant to social sustainability) have been advanced, for example about residential stability, social capital and better neighbourhood outcomes, particularly in education (Aaronson, 2000; Haurin *et al.*, 2002; Dietz and Haurin, 2003; Hobcraft, 2002; Sigle-Rushton, 2004; Bramley and Karley, 2007), although the extent of these positive impacts on net social welfare has been questioned (Ford and Burrows, 1999; Rohe *et al.*, 2002). Other arguments have focused on asset-based welfare, namely the scope for growing wealth accumulation by older owners contributing to pension, care and other costs in old age (Doling and Ronald, 2010; Toussaint and Elsinga, 2009).

Whatever the merits and limitations of these arguments, it is clear that the actual trend of tenure in the UK over at least a decade has been in the opposite direction, namely a decline in owner occupation and a rise in private renting. Between 2001 and 2013 owner occupation declined from 69.5 per cent to 65.5 per cent of all households in the UK, while private renting rose from 7.4 per cent to 18.3 per cent in the same period; social renting continued a gradual decline from 18.9 per cent to 16.3 per cent. Most of the decline occurred between 2006 and 2010, suggesting that the GFC played a role, but it is also clear that owner occupation was flatlining from 2001 to 2005. If one looks at tenure by age cohort over time, it is clear that the early 'baby boomer' generation (those born between 1945 and 1955) have achieved the highest level of home ownership by middle age, and that each succeeding cohort seems to be on a lower level. This has given rise to the phrase 'generation rent'. Governments and political parties have responded to this with various policy initiatives. For example, in the 2015 general election the parties competed in offering policies to promote home ownership at the margins, including the ongoing 'help-to-buy' scheme (a shared-equity low-cost home ownership scheme), a related subsidised savings product, a shallow discount 'starter homes' scheme, and (most controversially) attempts to revive the 'right to buy' through larger discounts and extending it to the nominally independent third sector housing associations.

The fast-growing tenure of private renting offers attractive investment opportunities for wealthier older households concerned about their pensions, but it is ill-equipped to deliver a range of wider housing or social policy goals, such as tenure security, incentives to improve (particularly in respect of energy efficiency), stabilising populations in regeneration areas, controlling the cost of housing allowances, or a widened spread of asset-holding across the population.

Housing need trends

Table 1.2 suggests that the UK has an intermediate position in terms of key measures of housing problems/outcomes, which may be referred to as 'housing needs', when compared with other EU countries. Overall, the UK lies rather above the average for EU15 countries in terms of a range of needs identified, but generally below the EU28 level. We can turn to different sources (UK longitudinal panel household surveys) for an analysis of housing need outcomes to consider change over time. In an earlier study (Bramley, 2015: Figure 1), I looked at five measures which are available and comparable for England over the whole period 1991–2011. For four of the five needs, the pattern over time is essentially U-shaped, with needs tending to fall in the 1990s, and in some cases up to the mid-2000s, then rising in the most recent period. For four of the five categories, the latest period appears to show the highest levels of need, although some caution is necessary about the precise comparability between the surveys. The exception is affordability problems; here it is relevant to note that 1991–1992 was the high point of an earlier 'affordability crisis' in British housing (Bramley, 1994). Some other need indicators cannot be compared consistently over such a long period. There was a strikingly larger fall in traditional house condition problems between 1996 and 2008, and some fall in unsuitability (health or family related), but some apparent rise in sharing or lacking amenities (Bramley, 2015: Figure 2). The latter could be attributable to the rise of private renting.

There is considerable interest in international comparative housing studies in the relationship between housing needs/deprivations and more general poverty status levels (Bramley and Besemer, 2013; Bradshaw *et al.*, 2008). I have suggested one way of looking at this relationship over time (Bramley, 2015: Figure 7). Households with any of the needs just described are categorised into those who could, on a conventional mortgage-type test, afford to buy a home of the relevant size in their local area, and those who could not. Two poverty measures are used: relative low income after housing costs; and a combination of this with material deprivation (Gordon and Pantazis, 1997; Guio, 2010).

This analysis shows that not only are the non-poor much less likely to experience need, but also that a lot of the needs they do experience could be solved by them moving to buy (although that proportion falls somewhat over time). The relative income-poor group are about three times more likely to have needs which they cannot move to avoid. The poor who lack material necessities have a significantly higher incidence of needs, which are mainly unavoidable, and these increased even in the earlier period, in contrast with the situation of the non-poor and the relative income poor. A majority of this group have had unavoidable housing needs since 2001.

This seems to provide quite strong evidence for the relationship between poverty and bad housing, even in the British case, where, it is often argued, the combination of large social sector, extensive rights for homeless households and comprehensive housing allowance tends to break that link (Bradshaw *et al.*, 2008). However, it should also be noted that there is a far from complete overlap between needs and poverty at the individual level.

Sustainable housing and neighbourhoods in Britain

This section looks in more depth at some aspects of the sustainability of housing and neighbourhoods, drawing on more specific and in-depth research using both established and new secondary data sources and some new primary data. The particular issues addressed are domestic energy and the social sustainability of neighbourhoods.

Energy efficiency and fuel poverty

One of the critical elements of housing system performance from an environmental point of view is the energy efficiency of the housing stock, given that domestic energy accounts for around a quarter of CO_2/greenhouse gas emissions. Although the UK inherited a large housing stock from earlier eras, those eras had cheaper fuel costs and generally lower building standards. There have been major investments in upgrading the social housing stock through the 'Decent Homes' programme from 2001 (and equivalents in Scotland, Wales and Northern Ireland), which have brought about a significant improvement in thermal efficiency. Various grant/incentive programmes have existed for private house owners over time, although these have recently tended to focus either on 'vulnerable' older occupiers with limited means, or to take the form of loan schemes. Figure 8.2 shows a pattern of gradual improvement over the last two decades, with the social sector outperforming the private sector, but also showing a continuing large gap between the existing and new stock. The lowest performance, and also probably the greatest gap in terms of policy measures, is the private rented sector. Here there is the additional problem that tenants pay the energy bills, and would therefore

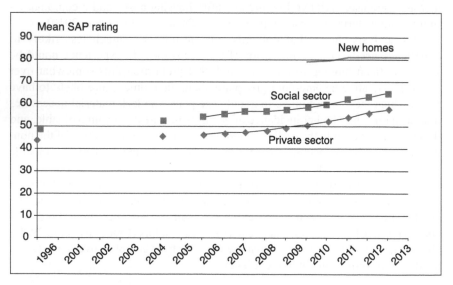

Figure 8.2 Energy efficiency measures for existing housing stock by tenure and for new homes.

Sources: ONS, 2014b: Figure 28.1; DCLG, 2015.

Note: SAP = Standard Assessment Procedure for energy rating of residential dwellings.

benefit from any savings, whereas landlords have to pay for any investment in upgrading energy efficiency.

According to the DCLG (2015), the majority of UK homes (16.3 million, or 70 per cent) could theoretically benefit from at least one energy improvement. The most common improvements would be more efficient boiler, cavity wall insulation or loft insulation. The average cost per home affected would be €1,265, but for some it would be as low as €410. Applying all indicated upgrades could reduce energy bills by 14 per cent, leading to a reduction in the total domestic energy bill from €29 billion to €25 billion. Applying all of the improvements recommended in Energy Performance Certificates would lead to a 22 per cent reduction in notional CO2 emissions, equivalent to 1.1 tonnes per dwelling per year.

A strand of policy has aimed to raise the standard for new homes to a much higher level, with the concept of 'zero-carbon homes' proposed as something to be achieved by a target date. Regulatory control of building standards has been used to try to achieve this, and the house-building industry has been encouraged to innovate and prepare for higher standards. However, in 2015, this policy was substantially abandoned, because of the apparent effects of the high costs of compliance on viability of new house-building post-recession. Nevertheless, it can be seen from Figure 8.2 that new homes are achieving relatively high standards, and improving gradually, although they are still far from 'zero carbon'.

In contrast with the countries of southern Europe, in the cold and damp UK the main issue is keeping warm rather than cooling, with air-conditioning still rare in domestic housing. However, with climate change becoming a reality, potential overheating of some UK housing types, especially in the south of England, may become more of an issue (Williams *et al.*, 2012).

Domestic energy efficiency has a dual motive in the UK – to reduce carbon emissions and to alleviate 'fuel poverty'. The latter is a formally recognised problem, referring to the situation where households are pushed into (deeper) poverty because of the low thermal efficiency of their homes in combination with the price of energy. Ways of measuring it now vary, following Hills (2012), but generally there was a big increase in the problem following energy price rises in the later 2000s. Evidence from the UK's Poverty and Social Exclusion surveys (Lansley and Mack, 2015; Bramley and Besemer, 2013) suggests that several of the forms of deprivation which worsened between 1999 and 2012 were due to this.

Within the private market there have been various attempts to try to raise consumers/home-buyers' awareness of energy efficiency, for example through information requirements needed for house transactions, including an Energy Performance Certificate. There is some evidence that more energy efficient homes command some price premium (Dunse *et al.* 2011; Thanos and Dunse, 2012). The latest scheme targeted at home owners, the 'Green Deal', seems to have been relatively unsuccessful, however, perhaps because it was designed to be run mainly by private contractors and finance. There is a further difficulty in reaching the increasingly important private rented sector – the mismatched incentives problem. Here there is now a proposal to have a mandatory minimum energy efficiency standard.

Social sustainability, urban form and social mix

As noted earlier, there has been growing interest in the social dimension of sustainability and how this may be related to aspects of planning and housing policy, particularly urban form and social mix. Lack of space prevents a full review of the relevant literature, but it is worth referring to our own contributions to this, particularly the 'CityForm' research (Bramley and Power, 2009; Bramley *et al.*, 2009; Dempsey *et al.*, 2012) and a more recent conference paper based on analysis of the first wave of the Understanding Society survey for England (Bramley and Watkins, 2013). These papers showed that, based on analysis of household survey evidence linked to small area urban form characteristics, there was a general pattern whereby outcomes relating to social equity, particularly access to services and opportunities, tended to be better in denser, more centrally located neighbourhoods, whereas outcomes relating to the sustainability of community (place attachment, stability, social interaction, safety, environmental quality) tended to be better in less dense, less centrally located neighbourhoods with more single-family housing rather than flats. There were some interesting nuances to this, such as the rather greater social interaction in neighbourhoods with intermediate levels of density, characterised by terraced (row) housing rather than detached housing (Dempsey *et al.*, 2012). It was also shown that the socio-economic status of households and areas had rather stronger influence on these outcomes, with poorer households and neighbourhoods having worse outcomes (Bramley and Power, 2009).

The Bramley and Watkins (2013) paper enabled a more nuanced analysis as well as the inclusion of health and wellbeing outcomes. Using a combination of correlation, factor and regression analyses, it was shown that outcome measures for the sustainability of community dimension could be grouped into two sub-dimensions of 'social capital', with one related more to attachment/stability and the other to community engagement, with health and wellbeing emerging as a third general factor. Urban form was represented by a general factor for distance/ green space, another for density, and more specific measures relating to large blocks, entrance on ground floor, and mixed use. It was found that the distance/ green space factor had consistent and strong positive effects on the social sustainability outcomes, whereas the density factor was less significant; large blocks were generally negative, ground-floor entrance positive, with mixed findings for mixed use (particularly negative for environmental quality). Previous findings (e.g. Bramley and Power, 2009) for the negative effects of low income/ status were broadly confirmed. Some findings suggested that mixing of socio-economic and ethnic groups could be negative, but there are ambiguities here and one should exercise caution about relying on subjective 'similarity' measures, which are poorly related to objective similarity measures. Relationships were found to be generally robust to the inclusion of additional intervening variables and attitudinal proxies for possible selection effects. An important positive finding was that social capital, particularly of the 'engagement' kind, had a positive effect on general wellbeing when controlling for everything else. This suggests that it is valid to seek to promote social capital as a route to generally happier urban communities.

Conclusions: challenges in policy implementation

Given the context described earlier in this chapter, it is clear that urban planning and housing policies in the UK face a number of challenges for implementation, if not outright contradictions. These are briefly highlighted below:

- *Demand management* – the pressures of demand on the housing market are high and difficult to satisfy, and this raises questions about whether and how key national policies should be developed to manage this, for example in respect of immigration, in future mortgage regulation, and in the use of demand-side subsidies such as the measures recently introduced to support first-time buyers.
- *Localised planning* is popular and may be a means to ensure that new developments are more sensitive to issues of social sustainability. Nevertheless, it a major obstacle to housing supply delivery, particularly in the high-demand areas of the south of England. This includes the politics of the green belt, and the difficulty of coordinating local authority plans in interconnected sub-regional markets without a regional planning machinery.
- *Quantity versus quality* – the part of the planning system that is intended to promote economic growth and housing supply, including the 'presumption in favour of sustainable development' leading to major appeal decisions in the many areas which do not have up-to-date, agreed and adequate local plans, could lead to poorly located, poorly serviced housing developments which repeat past mistakes in terms of unsustainable communities. These dangers are reinforced by the reduced resources for local authority planning consequent upon fiscal austerity. More positively, better-quality development which contributes to social sustainability, wellbeing and happiness could also make communities take a more positive view towards new house-building.
- *London* – the pattern of demand and development across the regions is very skewed, with London exerting an over-dominant influence, with insatiable excess demand, while northern industrial areas lag behind, are hard hit by public finance austerity, and lack the resources to address urban regeneration issues. Within London, the property market and welfare cuts are leading to rampant gentrification and the progressive expulsion of the poor from the centre. There are pressures to adopt super-density solutions of dubious social and environmental sustainability, while the green belt is treated as sacrosanct and opportunities for well-planned urban extensions are thereby frustrated. Experience has shown that such high land values can be recaptured for community benefit, in the form of affordable housing or community amenities, through planning obligations, but this does require a consistently supportive policy framework from national government.

- *Mixed communities* – a previous consensus for social mix, implemented through planning obligations ('Section 106'), is being weakened by short-term 'quick fixes' for supply following the recession, while vote-catching 'right to buy' extensions are to be funded by selling off social housing in high-value areas and threaten to undermine the foundations of the housing association sector.
- *Housing Benefit/Allowance* – this distinctive and important system underpinning affordability and access for low-income groups is threatened by the probability that it is financially unsustainable and a likely target for further welfare reforms and cuts. This will mean that more lower-income households will be exposed to market forces, with increasing vulnerability to a range of housing deprivations and housing-induced poverty.
- *Intergenerational wealth issues* – there is a growing divide between a large 'comfortable' class of older home owners with substantial wealth and secure pensions, untouched by welfare reforms because of their voting power, and a large group of younger, working-age households facing diminishing prospects of stepping on the 'housing ladder', including a large and growing sector in 'precarious' employment. While there will be some spontaneous shift towards more reliance on family financial transfers across the generations in response to this, making the UK a little more like the Mediterranean countries, this process will largely replicate and reinforce wealth inequalities across the generations.

References

Aaronson, D. (2000) 'A note on the benefits of homeownership', *Journal of Urban Economics*, 47: 356–369.

Barker, K. (2004) *Review of Housing Supply: Delivering Stability: Securing our Future Housing Needs.* Final report and recommendations. London: TSO/H. M. Treasury.

Barton, H., Grant, M. and Guise, R. (2003) *Shaping Neighbourhoods: A Guide for Health, Sustainability and Vitality.* London: Spon Press.

Bozio, A., Emerson, C., Peichl, A. and Tetlow, G. (2015) 'European public finances and the Great Recession: France, Germany, Ireland, Italy, Spain and the United Kingdom compared', *Fiscal Studies*, 36(4): 405–430.

Bradshaw, J., Chzehn, Y. and Stephens, M. (2008) 'Housing: the saving grace in the British welfare state?', in Fitzpatrick, S. and Stephens, M. (eds) *The Future of Social Housing.* London: Shelter.

Bramley, G. (1994) 'An affordability crisis in British housing: dimensions, causes and policy impact', *Housing Studies*, 9(1): 103–124.

Bramley, G. (1997) 'Housing policy: a case of terminal decline?', *Policy and Politics*, 25(4) 387–407.

Bramley, G. (2007) 'The sudden rediscovery of housing supply as a key policy issue', *Housing Studies*, 22: 221–242.

Bramley, G. (2012a) 'Affordability, poverty and housing need: triangulating measures and standards', *Journal of Housing and the Built Environment*, 27(2): 133–151.

Bramley, G. (2012b) 'Localised planning, local sentiment and planning stances: evidence and likely outcomes for housing in England.' Paper presented at Policy and Politics Anniversary Conference, Bristol, September.

Bramley, G. (2013) 'Housing market models and planning', *Town Planning Review*, 84(1): 9–34. DOI: 10.3828/tpr.2013.2.

Bramley, G. (2015) 'Housing need outcomes in England through changing times: demographic, market and policy drivers of change', *Housing Studies*, 31(3): 246–268.

Bramley, G. and Besemer, K. (2013) 'The role of housing in the impoverishment of Britain'. Paper presented at European Network for Housing Research Conference, Tarragona, June.

Bramley, G., Brown, C., Dempsey, N., Power, S. and Watkins, D. (2009) 'Urban form and social sustainability: evidence from British cities', *Environment and Planning A*, 41(9): 2125–2142.

Bramley, G. and Karley, N. K. (2007) 'Homeownership, poverty and educational achievement: school effects as neighbourhood effects', *Housing Studies*, 22(5):693–721.

Bramley, G. and Power, S. (2009) 'Urban form and social sustainability: the role of density and housing type', *Environment and Planning B: Planning and Design*, 36(1): 30–48.

Bramley, G. and Watkins, D. (2013) 'Social sustainability and urban form in harder times'. Paper presented at the European Network for Housing Research Conference, Tarragona, June.

Bramley, G. and Watkins, D. (2014) 'Coalition of the willing: the shape of urban growth in England post-crisis'. Paper presented at the European Network for Housing Research Conference, Edinburgh, July.

Coleman, J. S. (1988) 'Social capital in the creation of human capital', *American Journal of Sociology*, 94: S95–S120.

Department for Communities and Local Government (DCLG) (2012) *National Planning Policy Framework*. London: DCLG.

Department for Communities and Local Government (DCLG) (2015) *English Housing Survey: Energy Efficiency of English Housing*. London: DCLG. Available at www.gov. uk/government/uploads/system/uploads/attachment_data/file/445440/EHS_Energy_ efficiency_of_English_housing_2013.pdf, accessed 11 July 2016.

Department for Communities and Local Government (DCLG) (n.d.) *Live Tables on Housing Market and House Prices*. Available at www.gov.uk/government/statistical-data-sets/live-tables-on-housing-market-and-house-prices#live-tables, accessed 8 August 2016.

Department of Environment, Food and Rural Affairs (DEFRA) (2005) *The UK Sustainable Development Strategy*. London: HMSO.

Department of the Environment, Transport and Regions (DETR) (1997) *Indicators of Sustainable Development*. London: DETR.

Department of the Environment, Transport and Regions (DETR) (2001) *Achieving a Better Quality of Life: Review of Progress towards Sustainable Development*. London: HMSO.

Dempsey, N., Bramley, G., Power, S. and Brown, C. (2009) 'The social dimension of sustainable development: defining urban social sustainability', *Sustainable Development*, 19(5): 289–300.

Dempsey, N., Brown, C. and Bramley, G. (2012) 'The key to sustainable urban development in UK cities? The influence of density on social sustainability', *Progress in Planning*, 77(3): 89–141. Available at http://dx.doi.org/10.1016/j.progress.2012.01.001, accessed 11 July 2016.

Dietz, R. and Haurin, D. (2003) 'The social and private micro-level consequences of homeownership', *Journal of Urban Economics*, 54: 401–450.

Doling, J. and Ronald, R. (2010) 'Home-ownership and asset-based welfare', *Journal of Housing and the Built Environment*, 25: 165–173.

Dunse, N., Thanos, S. and Bramley, G. (2011) 'Relationships between house value, energy expenditure and energy ratings'. Paper presented at the 18th European Real Estate Society Conference, Eindhoven, June. Available at http://discovery.ucl.ac.uk/1411174, accessed 7 August 2016.

Fitzpatrick, S., Pawson, H., Bramley, G., Wilcox, S. and Watts, B. (2015) *The Homelessness Monitor: England 2015*. London: CRISIS. Available at www.crisis.org.uk/data/files/publications/Homelessness_Monitor_England_2015_final_web.pdf, accessed 7 August 2016.

Ford, J. and Burrows, R. (1999) 'The costs of unsustainable home ownership in Britain', *Journal of Social Policy*, 28: 305–330.

Forrest, R. and Kearns, A. (2001) 'Social cohesion, social capital and the neighbourhood', *Urban Studies*, 38(12): 2125–2143.

Freeman, L. (2001) 'The effects of sprawl on neighbourhood social ties', *Journal of the American Planning Association*, 67(1): 69–77.

Fukuyama, F. (2000) *The Great Disruption: Human Nature and the Reconstitution of Social Order.* London: Profile Books Ltd.

Gibb, K (2014) 'The multiple policy failures of the UK bedroom tax'. *International Journal of Housing Policy*, 15(2). Available at www.tandfonline.com/doi/full/10.1080/14616718.2014.992681, accessed 7 August 2016.

Glynn, T. (1981) Psychological sense of community: measurement and application', *Human Relations*, 34: 789–818.

Gordon, D. and Pantazis, C. (1997) *Breadline Britain in the 1990s*. Aldershot: Ashgate.

Guio, A.C. (2010) 'EU 2020 poverty reduction: concepts, figures and challenge'. Paper presented at Countdown to Europe 2020 Conference, Centre for Parliamentary Studies, Brussels, 29 June.

Guiu, A.-C., Gordon, D. and Marlier, E. (2012) *Measuring Material Deprivation in the EU: Indicators for the Whole Population and Child-Specific Indicators*. Report presented to EU Task Force on Material Deprivation.

Haughton, G. and Hunter, C. (1994) *Sustainable Cities*. London: Jessica Kingsley.

Haurin, D.R., Parcel, T.L. and Haurin, R. J. (2002) 'Does homeownership affect child outcomes?', *Real Estate Economics*, 30: 635–666.

Hills, J. (2012) *Getting the Measure of Fuel Poverty: Final Report of the Fuel Poverty Review*. CASE Report 72. London: London School of Economics.

Hills, J. (2015) *Good Times, Bad Times: The Welfare Myth of 'Them' and 'Us'.* Bristol: Policy Press.

Hills, J., Le Grand, J. and Piachaud, D. (2002) *Understanding Social Exclusion*. Oxford: Oxford University Press.

Hirschfield, A. and Bowers, K. (1997) 'The effects of social cohesion on levels of recorded crime in disadvantaged areas', *Urban Studies*, 34(8): 1275–1295.

Hobcraft, J. (2002) 'Social exclusion and the generations', in Hills, J., Le Grand, J. and Piachaud, D. (eds) *Understanding Social Exclusion*. Oxford: Oxford University Press.

IMF (International Monetary Fund) (2015) *World Economic Outlook Database: Historical WEO Forecasts Database: April 2015*. Available at www.imf.org/external/pubs/ft/weo/2015/01/weodata/index.aspx, accessed 7 August 2016.

Independent (2015) 'Gentrification pushing some of the poorest members of society out of their homes, says study', 15 October.

Jenks, M., Burton, E. and Williams, K. (1996) *The Compact City: A Sustainable Urban Form?* London: E. and F. N. Spon.

Lansley, S. and Mack, J. (2015) *Breadline Britain: The Rise of Mass Poverty*. London: OneWorld.

Layard, R. (2005) *Happiness: Lessons from a New Science*. London: Penguin.

Leishman, C. (2015) 'Housing supply and suppliers: are the microeconomics of housing development important?' *Housing Studies*, 30(4): 580–600.

Mitlin, D. and Satterthwaite, D. (1996) 'Sustainable development and cities', in Pugh, C. (ed.) *Sustainability, the Environment and Urbanization*. London: Earthscan.

Nash, V. and Christie, I. (2003) *Making Sense of Community*. London: Institute for Public Policy Research.

Office of the Deputy Prime Minister (ODPM) (2003) *Sustainable Communities: Building for the Future*. London: HMSO.

Office of National Statistics (ONS) (2014a) *Annual Mid-year Population Estimates, 2013*. Available at www.ons.gov.uk/peoplepopulationandcommunity/populationandmigration/populationestimates/bulletins/annualmidyearpopulationestimates/2014-06-26, accessed 11 July 2016.

Office of National Statistics (ONS) (2014b) *Sustainability Indicators*. Available at http://webarchive.nationalarchives.gov.uk/20160105160709/http://ons.gov.uk/ons/rel/wellbeing/sustainable-development-indicators/july-2014/index.html, accessed 11 July 2016

Pantazis, C., Gordon, D. and Levitas, R. (2006) *Poverty and Social Exclusion in Britain: The Millennium Survey*. Bristol: Policy Press.

Pawson, H. and Davidson, E. (2008) 'Radically divergent? Homelessness policy and practice in post-devolution Scotland', *International Journal of Housing Policy*, 8(1): 39–60.

Pawson, H. and Mullins, D. with Gilmour, T. (2010) *After Council Housing: Britain's New Social Landlords*. Basingstoke: Palgrave Macmillan.

Picketty, T. (2014) *Capital in the Twenty-first Century*. Translated by Arthur Goldhammer. Cambridge, MA, and London: Belknap/Harvard University Press.

Pierson, J. (2002) *Tackling Social Exclusion*. London: Routledge.

Putnam, R. D. (2000) *Bowling Alone*. New York: Simon and Schuster.

Randall, C. (2012) *Measuring National Well-being: Where We Live*. Office for National Statistics. Available at http://webarchive.nationalarchives.gov.uk/20160105160709/http://www.ons.gov.uk/ons/dcp171766_270690.pdf, accessed 11 July 2016.

Rohe, W.M., Van Zandt, S. and McCarthy, G. (2002) 'Low income homeownership: examining the unexamined goal', in Tighe, J. R. and Mueller, E. J. (eds) *The Affordable Housing Reader*. Abingdon: Routledge.

Scanlon, K., Whitehead, C. and Fernández Arrigoitia, M. (eds) (2014) *Social Housing in Europe*. London: Wiley-Blackwell.

Sigle-Rushton, W. (2004) *Intergenerational and Life-Course Transmission of Social Exclusion in the 1970 British Cohort Study*. Case Paper 78. London: London School of Economics.

Talen, E. (1999) 'Sense of community and neighbourhood form: an assessment of the social doctrine of new urbanism', *Urban Studies*, 36(8): 1361–1379.

Thanos, S. and Dunse, N. (2012) *The Changing Effects on Domestic Energy Expenditure from Housing Characteristics and the Recent Rapid Energy Price Movements*. RICS Report. Available at www.rics.org/uk/knowledge/research/research-reports/domestic-energy-expenditure/, accessed 8 August 2016.

Toussaint, J. and Elsinga, M. (2009) 'Exploring "housing asset-based welfare": can the UK be held up as an example for Europe?', *Housing Studies*, 24(5): 669–692.

Urban Task Force (1999) *Towards an Urban Renaissance.* London: E. and F. Spon.

Whitehead, C., Scanlon, K. and Lund, J. (2014) *The Impact of the Financial Crisis on European Housing Systems: A Review.* SIEPS Report No. 2. Stockholm: Swedish Institute for European Policy Studies.

Wilkinson R. and Marmot, M. (eds) (2003) *Social Determinants of Health: The Solid Facts.* Copenhagen: World Health Organisation.

Williams, K. (2000) 'Does intensifying cities make them more sustainable?', in Williams, K., Burton, E. and Jenks, M. (eds) *Achieving Sustainable Urban Form.* London: E. and F. N. Spon.

Williams, K., Gupta, R., Smith, I., Joynt, J., Hopkins, D., Bramley, G., Payne, C., Gregg, M., Hambleton, R., Bates-Brkljac, N., Dunse, N. and Musslewhite, C. (2012) *Suburban Neighbourhood Adaptation for a Changing Climate (SNACC): Final Report.* Bristol: University of the West of England.

Wilson, W.J. and Taub, R. P. (2006) *There Goes the Neighborhood: Racial, Ethnic and Class Tensions in Four Chicago Neighborhoods and Their Meaning for America.* New York: Alfred A. Knopf.

World Commission on Environmental Development (1987) *Our Common Future.* Oxford: Oxford University Press.

9 Romania

Catalina Turcu

Introduction

This chapter focuses on the second-largest country in Eastern Europe after Poland, Romania. An EU member since 2007, Romania has a population of 20 million (see Table 1.1) and has been undergoing a process of transition from a former centralised economy to a market economy since the fall of the communist regime in 1989. Romania's population is declining and aging due to low birth rates and outward migration. It decreased by about 1.7 million people between 1990 and 2007, the equivalent of a 7.2 per cent population loss (GovRo and UNDP, 2008). At this pace, it has been estimated that Romania's population will plummet from 20.8 million in 2013 to 15 million by 2050 (GovRo and UNDP, 2008). This will have major implications for the overall size of the labour force and, thus, Romania's economic outlook.

Despite these negative demographic trends, Romania saw notable improvements in its macroeconomic performance between 2001 and 2007, with an average annual GDP growth of 6.7 per cent, compared to countries such as Germany, with 1.5 per cent (Eurostat, 2015). This was one of the highest economic growth rates in Europe and occurred alongside a relatively successful process of macro-stabilisation (GovRo and UNDP, 2008). Romania's GDP per capita (€7,200) is, however, among the lowest in the EU at only 28 per cent of the EU28 average (€25,500 per capita). This is also significantly lower than countries such as Norway (€77,400 per capita) or even Hungary (€10,200 per capita) (see Table 1.1).

Romania entered recession in 2009, registering a sharp decline in industrial production, construction, exports and lending activity (Zaman and Georgescu, 2009). There have been cuts in public funding, but also cuts in research and development and environmental protection investment. The effects of the economic crisis have mainly been felt in urban areas, where investment in a previously booming real estate market ceased and many businesses closed down. Public wages have been cut, with Romania registering the highest cuts in public pay in Europe, at 25–50 per cent in 2010 (Grimshaw *et al.*, 2012). Pensions have also been frozen, and social assistance and public investment curtailed. These cuts have been imposed due to funding cuts from the EU and the IMF; €20 billion were granted to Romania in 2009, compared to €85 billion to Ireland in 2010, and €219 billion to Greece in 2010–2011 (Grimshaw *et al.*, 2012). However, the size of public sector employment in Romania is the smallest (below 15 per cent) among EU28 countries, and less than half that of Sweden (33 per cent).

The recent crisis has had a relatively smaller impact on Romania's overall economic outlook than those of countries such as Spain and the UK. Since 2011, Romania has started to recover steadily and its GDP per capita has recovered faster than in those countries. By contrast, the UK still registered zero growth in 2012 and Spain grew only slightly in 2014 (Eurostat, 2015). Moreover, Romania's current unemployment level (6.5 per cent) is relatively low – similar to that of the Netherlands, lower than Sweden's (7.7 per cent) and significantly lower than Spain's (24.1 per cent) (see Table 1.1).

Romania offers a range of welfare benefits, such as maternity entitlement, free healthcare, tuition-free education, special allowances for the disabled, and child benefit proportional to the number of children in the household. However, this welfare provision is seen as less generous than in Western Europe, a characteristic of the so-called 'transition' welfare regimes (Castles *et al.*, 2010). For example, unemployment benefits are among the lowest in Europe; pensioners have been largely ignored over recent decades; and single-parent households and ethnic minorities are little addressed by the current welfare system.

Romania has a centralised form of government and, in administrative terms, is split into approximately 3,000 local authorities (called '*comune*', '*orase*' and '*municipii*') grouped under 42 counties (called '*judete*'). There is no legally recognised regional level in Romania, although the country has eight geographic regions. The national government issues legislation, local authorities have the duty to implement it, and counties 'police' implementation of the legislation at the local level (EU, 2015). Local authorities deal with social housing, social assistance and services, urban planning and urbanism, urban infrastructure and so on. Therefore, Romania has two active levels of government: the national or legislative level and the local authority or operational level.

Since 1990, Romania's territory has undergone a relatively chaotic process of urbanisation and spatial development against the background of the transition from a heavily ideological and centralised planning system to a still centralised but more market-responsive one. During this time, Romania has seen the erosion of its cultural heritage, significant urban sprawl and the socio-economic decline of its mono-industrial cities and towns. Today, its urban areas are confronted with a dated urban infrastructure and public realm, a significant amount of substandard housing, traffic congestion, high levels of pollution, top-heavy city management structures and significant sprawling supported by a market-driven planning system (GovRo and UNDP, 2008).

This chapter provides a brief overview of Romania's housing system and policy, followed by an examination of its approach to sustainable communities. It then moves on to draw on the author's primary research on the energy retro-fitting of apartment building housing in Bucharest. Finally, it reflects on whether Romania's current energy retrofitting initiatives can be seen as a means to more sustainable urban housing and communities.

The housing system in Romania

Romania's housing stock is old, with 90 per cent built before 1990. Sixty-five per cent of it was built during the socialist period (1950–1990) and more than half (53 per cent) before 1970, when planning regulations and building standards were less stringent (Figure 9.1). Buildings built between 1970 and 1990 (37 per cent of the total stock) have already reached their thirty-plus life-cycle milestone, when capital renovation, adaptation and upgrades are needed. Moreover, many of the buildings which have reached their fifty-plus threshold (built between 1950–1970; 28 per cent of the total stock) did not benefit from major reinvestment at thirty-plus (in the late 1990s and early 2000s) and are of poor quality and long overdue capital investment.

In addition to its poor condition and old age, Romanian housing is currently the most overcrowded in Europe, with more than half (51.6 per cent) classified by Eurostat as 'overcrowded housing', compared to only 17 per cent in EU28 (see Table 1.2). Romania also ranks as the lowest of the EU28 countries in terms of housing quality at 56.5 per cent (see Table 1.2). One in five Romanians suffered from severe housing deprivation in 2014 – that is, living in overcrowded housing and lacking a bath/toilet, or having a leaky roof, or living in accommodation that was considered too dark (Eurostat, 2016a).

Sixty per cent of Romanians live in single houses, one of the highest proportions of detached-house living in Europe and significantly above the EU28 average of 34 per cent (Table 1.2). The rest of the population (37.8 per cent) live in flats or apartment buildings, mainly situated in large urban housing estates. Moreover, home ownership is the main housing tenure, with Romania known as a 'super home ownership' nation. Its home ownership rate is the highest in Europe at 96.6 per cent. This is a direct result of the mass privatisation of state-owned homes to the sitting tenants at extremely low prices (or even for free) in the early

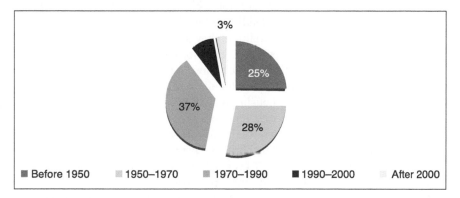

Figure 9.1 Housing stock in Romania by age of stock.

Source: Compiled by the author from MDRAP (2010) data.

1990s. Home ownership increased from 67.3 per cent in 1990 to over 93 per cent in 1993 and 96.6 per cent by 2012 (Eurostat-SILC, 2012).

The private rented sector is almost non-existent in official statistics, at just 0.8 per cent. However, experts suggest that this is a significant underestimate and the actual proportion is 11–15 per cent. The difference is explained by the existence of a 'black market' created by landlords to avoid paying taxes (Bejan *et al.*, 2014). The sector is unregulated and, apart from tax evasion, is hampered by tenure insecurity and unpredictable tenant–landlord relations due to the lack of contractual agreements and affordability problems. This is especially a problem in urban areas. For example, the housing cost overburden rate stands at 76.3 per cent for private tenants (see Table 1.2).

The proportion of social housing or publicly rented housing is also small, estimated at 2.3–2.6 per cent of the total stock (Housing Europe, 2015). Social housing is financed by national government and built, owned, allocated and maintained by local authorities, the only type of social landlord in Romania. It is regarded as housing for the poorest groups in society and is a residual sector today due to the housing privatisation of the early 1990s and subsequent home ownership programmes. The sector is dominated by 'sitting since the 1990s' and Roma tenants (Bejan *et al.*, 2014).

The Housing Act (Law No. 114/1996) regulates social housing in Romania. It states the income threshold for eligible social tenants – that is, average monthly income per person needs to be below the national average (€217 in 2015) – and limits the amount of rent paid to less than 10 per cent of the household's monthly income. The Act also defines the categories who are entitled to social housing: households rendered homeless by natural disasters (such as earthquakes or flooding) and property retrocession processes; adults under thirty-five who cannot afford to buy housing at market price; adults leaving social care institutions; the disabled; pensioners; veterans and widows of the Second World War; political dissidents from the previous socialist regime; and the families of martyr-heroes who participated in the Romanian Revolution of December 1989.

The Act also places a 'duty of care' for social housing with local authorities. They are responsible for rent subsidies, allocation and maintenance of social housing in their jurisdictions. In addition, they are free to prioritise allocation as they see fit and secure finance for repairs from local budgets or loans. Thus, it has been argued that local authorities see social housing as a financial burden because of their duty to subsidise rents and undertake repairs (Cosima-Rughinis, 2004). A very small amount of social housing is built every year. This is explained by social housing not being a priority for Romanian social policy but also by the sharp drop in funding for publicly built housing since 1989. The funding, as a share of public expenditure, declined from 8.7 per cent in 1989 to just 0.8 per cent in 2000 (Cosima-Rughinis, 2004). To sum up, the Romanian social housing stock is very small in size and targets extremely poor people.

Romania's current housing system has been shaped by the legacy of previous socialist policies, but also by rapid changes since 1989. Socialist housing policy was predominantly driven by state-provided housing, which built apartments to rent or buy on favourable terms. This saw the construction of poor-quality mass housing between the 1960s and the 1980s, which housed people near new industries during the country's rapid industrialisation. The allocation of state housing during this period promoted a socio-economic mix which was seen as an important tool for building an equal society (Kahrik and Tammaru, 2010). However, socialist housing allocation was not a completely fair and equal process. Certain groups (young married couples, families with children and university graduates) were favoured over others (the unemployed, unmarried couples and single people) (Turnock, 1990). Hence, it was seriously biased towards 'higher social positions', usually associated with higher education, occupation or party affiliation (Szelenyi, 1993; Turnock, 1990).

Some argue that a more balanced socio-economic mix was achieved by socialist state housing allocation in Eastern European countries than in some of their Western counterparts (Kahrik and Tammaru, 2010, Turkington *et al.*, 2004; van Kempen, 2005). For example, in the UK, state-allocated housing has mainly targeted disadvantaged groups (people on low incomes, single parents, the disabled, ethnic minorities) and is associated today with urban poverty and deprivation, social polarisation and crime (Hills, 2007). By comparison, socialist housing in Eastern Europe still displays fewer signs of socio-economic decline, higher proportions of better-educated residents and lower levels of deprivation than in Western Europe (Kahrik and Tammaru, 2010; Turkington *et al.*, 2004; van Kempen, 2005).

Housing policy in Romania has undergone two main developments since 1989: the 'residualisation' of social housing as a result of mass privatisation; and overt encouragement of private home ownership. Privatisation of state-owned housing in Eastern Europe has had three main aims (Clapham *et al.*, 1996). First, post-socialist states sought to reduce public expenditure and raise income from sales. Second, there was a genuine belief that markets would promote better housing provision than the state. Third, it was assumed that home ownership would break the 'state dependency culture' and 'encourage individuals to take more responsibility for themselves and their families' (Clapham *et al.*, 1996: 9). Homes in Romania were sold at heavily discounted prices to sitting tenants: in 1993, a one-bedroomed flat in certain areas of Bucharest could be purchased for the equivalent of a single average monthly salary of $100 (*Adevarul*, 2010).

There is no coherent framework for national housing policy in Romania today. Housing is the responsibility of the Ministry of Regional Development and Public Administration (Ministerul de Dezvoltare Regionala si Administratie Publica), which was created in 2009 through the merger of two ministries: the Ministry of Regional Development and Housing and the Ministry of Tourism. Under the new ministry, housing sits alongside other areas, such as planning and urbanism, thermal rehabilitation, construction and sustainable development. However, three more recent housing developments are worth mentioning here.

First, the National Housing Agency (Agentia Nationala pentru Locuinte) was established in 1999 with a view to supporting young employed people who cannot afford to buy housing – estimated at 135,000 in 2015 (CEB, 2015). Second, the

National Programme for the Thermal Rehabilitation of Apartment Buildings (Programul National pentru Reabilitarea Termica a Blocurilor de Locuinte) was launched in 2002 to deal with the energy retrofitting of apartments built between 1950 and1990. Third, the First Home (Prima Casa) Programme was launched in 2009 to facilitate state-backed, highly subsidised loans to first-time buyers for purchases not exceeding €60,000.

The recent economic crisis has not prompted a 'housing crisis' in Romania of the nature and intensity seen in other European countries. Romania's house building and reliance on 'toxic' mortgages were low when the crisis hit, especially when compared to Spain or the UK. Nevertheless, the crisis's impact on the housing sector was twofold. First, the construction sector contracted by 14 per cent in 2009 as a consequence of lower and more careful investment (Zaman and Georgescu, 2009). Despite slow recovery since 2011, lack of new housing has pushed more people to rent privately or live with their families. In addition, the lack of regulation of private renting has left the sector in something of a wilderness, with steep price increases compared to real incomes. Second, the crisis has had little impact on the country's banking system due to a tight national monetary policy, but it has affected lending activity. Although Romanians are not prone to taking out loans or living in debt (Tache and Neesham, 2010), even fewer people than before the crisis have been able to afford mortgages due to falls in household incomes and tighter lending conditions. When combined with delays/inability to make mortgage payments due to unstable/shrinking incomes and loss of employment, this led to a sharp fall in house prices, which continued to decline up to the end of 2014, when the first modest increase since the crisis began was recorded (Eurostat, 2014).

Sustainable communities in Romania

Sustainable development is defined by two equivalent terms in Romania: '*dezvoltare durabila*' and '*dezvoltare sustenabila*'. These terms are used inter-changeably and have emerged as synonymous borrowings from different linguistic sources – French and English, respectively. Sustainability has only started to permeate mainstream policy since Romania's entry to the EU in 2007 (Fatol, 2013). Thus, putting into practice the principles of sustainable develop-ment is still in its early days and some argue that its official acceptance merely pays lip service to European requirements (Dobrescu *et al.*, 2011).

Romania's understanding of sustainability and sustainable communities mirrors that of the EU and addresses seven dimensions, which are reflected in its 2008 National Sustainable Development Strategy. These are:

* climate change and clean energy;
* sustainable transport;
* sustainable consumption and production;
* preservation and natural resources management;
* public health;

- social inclusion, demography and migration; and
- poverty and sustainable development challenges at the global level.

For the clarity of the argument, however, my definition of sustainable development (hence, sustainable communities) draws on 'the prism of sustainability' developed by Valentin and Spangenberg (1999), which brings together four pillars: environmental, economic, social and institutional. The seven dimensions above can be classified under these pillars and I shall refer back to this definition towards the end of this chapter.

One report (GovRo and UNDP, 2008) has suggested that the delivery of sustainable development on the ground has been slower in Romania than in other Eastern European countries, and it has mainly focused on the environmental aspects of sustainability with results achieved in the following sectors:

- *Urban regeneration*: public realm modernisation and pedestrianisation in Cluj and Bucharest; investment and renewal in historical centres; and park/green space upgrading – for example, 65,000 trees and shrubs have been planted in Bucharest since 2010 (Fatol, 2013).
- *Energy in buildings*: national initiatives, such as the Thermal Rehabilitation of Apartment Buildings (MDRAP, 2010); Green House (Casa Verde), which aims to encourage renewable energy production (EC, 2012); and upgrading of urban district heating networks.
- *Green building practice*: new buildings are increasingly BREEAM and LEED certified in big cities such as in Bucharest, Cluj-Napoca and Iasi; and efforts are being made towards nearly zero energy/carbon buildings (NZEBs).
- *Public transportation*: the five big Romanian cities all have bike-sharing schemes (Bucharest and Cluj-Napoca since 2010; Constanta, Timisoara and Iasi since 2012); they have upgraded their public transportation fleets; and Bucharest is currently extending its underground system.

Barriers, challenges and drivers for sustainable communities in Romania

There are three main barriers to sustainable urban development and sustainable urban housing and communities in Romania – namely, economic, social and institutional barriers. Besides the gap in *economic* performance between Eastern and Western countries, the current economic crisis has impacted further on Romania's economic development and imposed severe public spending cuts. This has seen a poorer population and reduced budgets at the local level, which in turn has determined a reshaping of country's priorities around what are viewed as more pressing and emerging issues, specifically economic development and poverty, rather than sustainable development (Turcu, 2016; Turcu and Tosics, 2015). *Social* barriers are associated with weaker social capital in Romania, predominantly seen as a legacy of the previous socialist regime (Howard, 2003; Rose, 2009; Tosics and Hegedüs, 2003), but also cultural preferences and behaviour. Research notes the importance of social capital and human behaviour in creating sustainable communities (Czischke *et al.*, 2015; Turcu, 2012a, 2012b; Turcu and Moloney, 2015). Moreover, recent research highlights *institutional* barriers in relation to some areas of sustainable housing, such as energy efficiency and building retrofitting. They include weak human and financial institutional capacity; lack of skills and training; and 'silo' thinking. In addition, Romanian institutions are top-heavy and dominated by government institutions, while the role of community and civil society institutions has been downplayed. This is reinforced by a generalised lack trust in institutions (Turcu, 2016).

My research has identified six main challenges for sustainable urban development generally, and for sustainable urban housing and communities specifically:

- decarbonisation;
- sustainable urban transportation;
- conservation and management of natural resources;
- urban vulnerability/poverty;
- institutional frameworks; and
- spatial disparities.

Decarbonisation refers to the reduction of energy consumption in buildings via energy retrofits and modernisation of district heating systems. While almost all EU member states experienced a reduction per household in electricity consumption over the period 2005 to 2011, Romania (alongside Bulgaria) recorded the highest increase – by almost one-quarter compared to 2005 levels (Eurostat, 2013). In addition, Romania's burden of long-term neglect and lack of investment in the building stock is reflected in low levels of urban retrofitting.

Sustainable urban transportation is hindered by a significant increase in private car ownership since 1990. Romania doubled this fleet between 1995 and 2008, albeit from a small base when compared to other EU countries, and the trend is continuing (IEA, 2010). In addition, the use of public transport (train and bus) has significantly decreased from 29.7 per cent in 2000 (Eurostat, 2016b) to 17.8 per cent in 2012 (see Table 1.2). This has led in turn to urban traffic congestion, car parking encroachment onto public and green areas, and declining usage of public transportation. Romanian cities have long-running and well-established public transport systems which consist of buses, trams, trolley-buses and maxi-taxis. However, they are in urgent need of investment and only big cities have benefited from some investment in their existing urban transportation infrastructure.

The *conservation and management of natural resources*, such as water and waste in buildings, represents a particularly significant challenge for Romania. In fact, it has the lowest score on the Environmental Performance Index (EPI) in Europe at 50.5 (see Table 1.1), compared to Switzerland at 87. This is a particularly problematic issue in the areas of water and sanitation, water resources and fisheries. In addition, despite significant progress towards more sustainable water management (Eurostat, 2013), there is an acute lack of household water management systems in both urban and rural areas, wastage due to water leaks, and no water metering in buildings is still the norm. Moreover, Romania's performance on waste recycling is the worst in the EU, with only 1 per cent of municipal waste recycled in 2012 (EC, 2012). Romania has also seen a significant loss of biodiversity due to its chaotic development process.

There are challenges associated with *social exclusion, poverty and vulnerability* at the urban level. Romania needs to reassess its social services, especially following cuts in the aftermath of the crisis, and needs to invest in social infrastructure, such as community centres and housing services. Despite a decrease in the number of households in poverty between 2008 and 2011 (Eurostat, 2013), Romania still has one of the highest rates of poverty and social exclusion in Europe at 42 per cent, while the EU28 average is 25 per cent (see Table 1.1). In addition, Romania (along with Bulgaria) has the highest urban–rural disparity in poverty rates: people living in less urbanised areas are twenty times poorer than those living in more urbanised areas (Eurostat, 2013).

Institutions have been undergoing tremendous change since 1989 but still represent a major challenge in Romania. The country has a tradition of 'no trust' in public and political institutions, a legacy of the previous socialist regime and ongoing corruption and struggle for power in its aftermath. The recent crisis has impacted further on Romanians' trust in their government, overall political system and public institutions. Romanians are at the low end of the trust scale, while the Benelux and Scandinavian countries are at the high end (OECD, 2013). In addition, local authorities in Romania have limited financial and human capacity to implement sustainability in practice. Tensions and an overall institutional imbalance exist between formal and top-down institutions above informal and bottom-up organisations (Turcu, 2016; Turcu and Persson, 2015).

There are significant *spatial disparities* between Romania's cities, but also between Romanian cities and other European cities. These disparities have been fostered by different economic and social paths of development, use of resources, and quality of environmental and urban infrastructure. All Romanian regions have a GDP per capita below 75 per cent of the EU average and all are eligible for financing from the EU Structural Instruments under the Convergence Objective (GovRo and UNDP, 2008). Disparities also exist between big and small/medium towns, especially in mono-industrial localities that have been affected by economic restructuring post-1989. Some of these towns also have poor physical access with a negative impact on local development. In addition, the risk of poverty is more prevalent in less urbanised areas: the percentage of the population at risk of poverty and social exclusion in less urbanised regions is exceptionally high at 55 per cent, compared to 42 per cent for the population as a whole (Eurostat, 2013).

Turning to drivers of more sustainable communities and sustainable housing, a particularly important one is the country's need to reduce energy consumption with the residential sector a main area of concern but also potential (Turcu, 2015). Romania aimed to cut its total energy consumption by 13.5 per cent by 2016, relative to the 2001–2005 period (ICEMENERG and ANRE, 2012). In 2010, the housing sector became the main consumer at the national level with a share of 36.5 per cent, compared with 23 per cent for transport and 31 per cent for industry (ICEMENERG and ANRE, 2012). It has been estimated that in the housing sector alone energy consumption could be reduced by 41.5 per cent via various energy efficiency measures, such as the retrofitting of post-socialist housing estates (EC, 2006).

Almost half of all residential buildings in Romania (48 per cent; 3.1 million dwellings) are apartment buildings, which average forty apartments per building, in four–ten-storey buildings. The structure is brick in the case of low buildings and reinforced concrete and prefabricated panels for high-rise buildings. Almost all (94 per cent) of these apartment buildings are in private ownership; 72 per cent are situated in urban areas; and over 50 per cent are connected to district heating networks, where 92 per cent of the energy is supplied by combined heat and power (CHP) systems (BPIE, 2014).

Apartment buildings are pepper-potted across cities and towns in Romania, but mainly located in large housing estates that draw on Le Corbusier's '*ville radieusse*' ideal. These were planned to house the socialist 'working class' close to industrial locations. Today, these housing estates rely on a relatively efficient public transportation system as well as administrative, education and health services that are in need of modernisation. Since 1989, they have also seen some land-use diversification and densification via change of use and urban infill to accommodate small- to medium-scale additional commercial and business uses. I suggest that all of this forms a good starting point for compact, diverse and sustainable urban neighbourhoods and communities.

Energy retrofitting of apartment buildings in Romania

This section focuses on the environmental dimension of sustainable housing and communities in the form of energy retrofitting (also known as thermal rehabilitation) of apartment buildings in large socialist housing estates built between 1950 and 1990. It examines whether Romania is delivering more sustainable housing and communities via the thermal rehabilitation of this type of housing stock. It draws on primary data collected since 2012 and examines in some detail Romania's National Programme for the Thermal Rehabilitation of Apartment Buildings. This programme aims to contribute towards the three axes of Romania's National Sustainable Development Strategy (2008), discussed in the previous section: physical rehabilitation; improved resident wellbeing; and enhancement of the local economy. It is also seen as a means to warmer homes, more affordable utility bills, and additional jobs in the construction and energy-efficiency sectors.

Apartment buildings on large housing estates built between 1950 and 1990 are a liability for Romania due to poor physical conditions and consequent poor energy performance. Of the total energy wasted, 15–20 per cent is lost through roofs; 20–25 per cent through external (panelled) building envelopes; 20–25 per cent through windows and doors; and 5–10 per cent through poorly insulated basements (MDRAP, 2010). Up to 25 per cent of energy could be saved by retrofitting these buildings and a full return on the initial investment achieved in five to ten years (EC, 2006; Rotariu, 2012). Heating bills could be reduced by 40 per cent; CO_2 emissions by 30–40 per cent; and new jobs would be created (Rotariu, 2012; MDRAP, 2010).

Since 2007, the programme has delivered energy-efficiency measures such as insulation of external walls, roofs and basement floors, door and window double glazing, and pipe insulation (MDRAP, 2010). Initially, national government, local authorities and condominium associations were equal partners in the scheme, with each responsible for a third of the total cost. However, since 2012, the national government has paid for 50 per cent of the programme, with the local authorities contributing 30 per cent and the condominium associations 20 per cent. Total building costs are standardised via government ordinance (HG 363/2010) (see Table 9.1).

Spatial and institutional imbalances

There is no clear overview of the scale of delivery of apartment building retrofitting in Romania, and performance is patchy and little documented at the city and local authority levels (Turcu, 2015, 2016). Anecdotal evidence suggests

Photo 9.1 High-rise and medium-rise social housing in Bucharest.

Photo 9.2 High-rise and medium-rise social housing in Bucharest.

Table 9.1 Standardised costs (euros) for the thermal rehabilitation of different types of apartments

Apartment type	Average area (m²)	Cost/m² €	Total cost rehabilitation €	Apartment owner pays (20 per cent) €
Studio ('1 *camera*')	37	–	2035	407
One bed ('2 *camere*')	52	55	2860	572
Two bed ('3 *camere*')	66	–	3630	726

Sources: Rotariu (2012) and MDRT (2012).

that delivery varies significantly across the country, with Bucharest, Brasov, Cluj-Napoca and Timisoara at the forefront of energy retrofitting action, while the rest of the country lags well behind (Ungureanu, 2014). For example, Bucharest is seen as the country's 'champion city' in terms of energy retrofitting, with approximately 25 per cent of its estimated 8,000 apartment buildings renovated to date, compared to just 5 per cent for the rest of the country (Turcu, 2016).

Bucharest consists of six local authorities (or '*sectoare*': S1 to S6). They have a radial disposition in relation to the city centre and vary in the number of apartment buildings earmarked for thermal rehabilitation. The distribution of apartment buildings earmarked for renovation, those renovated to date, and the funding allocated at local authority level varies across these six local authorities (see Table 9.2). For example, despite having the smallest stock of apartment buildings earmarked for energy retrofitting, S1 and S2, Bucharest's wealthiest local authorities, were allocated the largest share of public funding in 2009–2010. They had retrofitted half of their stock by 2014. By contrast, S6 and S4 are less wealthy and have the largest stocks of apartment buildings in need of energy retrofitting in Bucharest. They were allocated proportionally less public funding in 2009–2010 and had renovated only 12 per cent and 7 per cent, respectively, of their stock by 2014 (Turcu, 2016).

This highlights the unequal spatial distribution of the energy retrofitting action across Bucharest's six local authorities. Wealthier local authorities (and resident communities) with a smaller stock of apartment buildings in need of energy retrofitting 'capture' most of the public funding and so deliver retrofitting faster and more efficiently. This happens at the expense of less wealthy local authorities (and resident communities), which are allocated less public funding despite larger stocks earmarked for energy retrofitting. My research shows that this unequal distribution can be explained as a function of municipal 'wealth' and energy retrofit spending, leadership, but also weak institutional interaction among the institutions of energy retrofitting (Turcu, 2016).

Spatial disparities of energy retrofitting action are also likely to occur between cities at the national level. Bucharest, as the capital city, but also other large Romanian cities, such as Cluj-Napoca, Timisoara, Iasi and Constanta, are more likely to attract public funding and deliver energy retrofitting than smaller cities and towns, especially in industrial locations/regions that have been affected by economic restructuring since 1989. For example, Bucharest was allocated half of the country's public funding for housing retrofitting in 2009–2010 for its estimated 10 per cent share of the total number of apartment buildings (80,000) across the country (Turcu, 2016).

The institutional landscape of apartment building retrofitting in Romania is also more complex than portrayed by current policy. Three main types of institutions are identified in national policy documents: national government (setting the agenda and drafting legislation); local authorities (playing a central role in managing the programme on a daily basis); and condominium associations (the programme's beneficiaries). However, my research identified four additional types of institutions that are not directly acknowledged at present: consultants, auditors and designers; developers and contractors; civil society organisations; and professional organisations and city/local authority networks (Turcu, 2016).

My research also found that trust among the different institutions involved in the energy retrofitting of apartment buildings in Bucharest was weak or even non-existent. As such, partnerships between institutions for apartment building retrofitting were limited and formed only on the basis of legally binding contractual agreements, such as those between local authorities and resident associations, or local authorities and contractors delivering building works

Table 9.2 Bucharest statistics by local municipality: apartment buildings earmarked for energy retrofit; buildings retrofitted by 2014; municipal budgets in 2014; municipal budgets in 2014 relative to buildings earmarked for energy retrofit; and 2009–2010 nationally allocated funding

| | Bucharest local authorities | | | | | | Bucharest total | Romania total |
	S1	S2	S3	S4	S5	S6		
Buildings earmarked for energy retrofit (number of buildings)	1100	1110	1425	1600	1023*	2114	8372	80,000
Number of buildings retrofitted by 2014 (as per cent of municipal total)	628 (57%)	511 (46%)	309 (22%)	117 (7%)	215 (14%)	244 (12%)	2024 (26%)	4000 (5%)
Municipal budget in 2014 (million euro)**	267	198	152	106	n/a***	220	–	–
Municipal budget in 2014/Buildings earmarked for energy retrofit	1.1	0.8	0.5	0.3	n/a***	0.5	–	–
Nationally allocated funding 2009–2010 in million euro (as per cent of Bucharest funding)	13.3	14	5	1.3	7.7	9.8	51.1	115

Sources: Compiled by the author from data in Vrabie (2014) and Ungureanu (2014).

Notes: * No statistics were available for S5, so the current value has been calculated by deduction from Bucharest total; ** Budget figures were compiled by the author from each individual municipality's website; *** S5 budget was not publicly available at the time of publication.

(Turcu, 2016). This can be seen as problematic, as previous reports have highlighted the importance of partnering and its role in delivering energy efficiency in building targets across Eastern Europe (BPIE, 2011; Klinkenberg Consultants, 2010).

So, is Romania delivering more sustainable housing via the energy retrofitting of its apartment building stock? My answer is 'yes', but to varying degrees when the definition of sustainable housing introduced above is considered (i.e. economic plus environmental plus social plus institutional factors). On the one hand, the thermal rehabilitation of apartment buildings addresses *environmental* (i.e. energy performance, energy consumption, CO2 emissions) and *economic* (i.e. energy costs for households, heating subsidies for local authorities) aspects of housing. The current focus is on improving energy performance from a pre-renovation annual energy consumption of 150–400kWh/m^2 to a post-renovation energy consumption of 100kWh/m^2 (BPIE, 2012). Evidence suggests that this is being achieved; and, overall, the residential sector is slowly becoming more energy efficient. In addition, Energy Performance Certificates (EPCs) have been introduced since 2013; and more energy advice and audits are now provided by local authorities (Energy Efficiency Watch, 2013).

On the other hand, *institutional* and *social* aspects of housing are largely neglected within current initiatives for energy retrofitting. My research shows that progress on the ground is hindered by tensions and lack of coordination between institutions. This leads to piecemeal delivery, but also misses an opportunity for integration with complementary initiatives, such as the upgrading of district heating networks, renewable energy production (under Casa Verde), water metering and waste recycling. Social aspects are also only marginally considered. Delivering warmer homes and reducing fuel poverty via cheaper electricity bills are certainly important social outcomes of thermal rehabilitation. However, resident involvement is a box-ticking exercise and residents are passive recipients throughout the delivery of works and in their aftermath. Thus far, local authorities have not encouraged capacity-building or behavioural change around energy use in apartment buildings (Turcu, 2016).

Conclusions

To sum up, Romania has just started to work on the sustainability of its urban housing stock, and economic, environmental, social and institutional aspects of sustainability are addressed to various degrees. This is a continuous and uphill battle as Romania has the worst Environmental Performance Index (EPI) score in Europe (see Table 1.1). Despite sluggish progress on the ground, however, changes are being made and urban communities are becoming slowly more sustainable. So, when the energy retrofitting of apartment buildings is considered, is Romania delivering more sustainable communities? My answer, once again, is 'yes', but to varying degrees.

The energy retrofitting of apartment buildings has a strong emphasis on the environmental and economic aspects of communities occupying this type of housing. Homes are warmer, less damp and cheaper to heat. This results in overall better quality of life as residents are likely to be healthier and better off. It also contributes to carbon savings and reduced air pollution. For example, it has been estimated that carbon emissions from heating, cooling, hot water and lighting in buildings can be reduced by between 71 and 90 per cent, depending on the rate of decarbonisation of the energy system (BPIE, 2011). Yet, one might argue that the current carbon targets ($100kWh/m^2/year$) are not ambitious enough, especially when compared to current passive house standards ($15kWh/m^2/year$). Because of this, a substantial part of the potential energy saving for this type of housing can be seen as 'locked in'.

The energy retrofitting of apartment buildings also has a series of economic benefits for urban communities, such as reduced energy bills (for local and national government as well as residents), further growth of particular economic sectors, GDP or public finances and, arguably, higher property values. It has been estimated that renovating Europe's buildings between now and 2050 could save as much as €1,300 million on energy bills (BPIE, 2011). Furthermore, increasing overall energy efficiency by 20 per cent might boost the EU's GDP by €33.8 billion by 2020 (EC, 2011). Investment in building retrofitting could also lift public budgets in member states by 0.5–1.0 per cent of GDP (Copenhagen Economics, 2012).

However, some important social and institutional aspects of sustainable communities have been neglected. They include community capacity-building and 'control' around energy retrofitting action and energy use in apartment buildings, but also wider institutional trust. For example, the current national programme is framed from the narrow perspective of carbon and cost reduction. In so doing, it fails to advocate and capitalise (with urban residents and communities) on the wider societal benefits of housing retrofitting, such as improved living conditions, less fuel poverty and better public health. At the same time, enhanced institutional trust can boost institutional interaction and delivery on the ground by building bridges among complementary initiatives, such housing energy retrofitting, district heating upgrading and sustainable transportation.

Last but not least, my research shows the uneven spread of energy retrofitting action across local authorities and, potentially, across the country. Wealthier local authorities (and economically stronger cities) receive most of the public subsidies, despite less need (i.e. smaller stocks of apartment buildings in need of energy retrofitting). This means that the current energy retrofitting initiatives do not reach the poorer local authorities (and economically weaker cities) with the highest need (i.e. the larger stocks of apartment buildings) and so the greatest potential for energy improvement. In other words, some communities are more likely to benefit from energy retrofitting action than others. This type of inequality raises questions of 'where' and 'for whom' apartment building retrofitting is delivered and highlights that certain communities can be 'excluded' despite their needs and because of geographic location or the wealth of their local authority. This new form of exclusion – 'exclusion from energy retrofit' – might well be found in other Eastern European countries and raises questions about the fairness and equality of current national initiatives for the energy retrofitting of apartment buildings in the region.

References

Adevarul (2010) 'România, tara proprietarilor de apartamente', 22 November. Available at http://adevarul.ro/news/societate/romania-tara-proprietarilor-apartamente-1_50 aecab47c42d5a663a07ab6/index.html, accessed 12 July 2016.

Arts, W. and Gelissen, J. (2010) 'Models of the welfare state', in Castles, F. G., Leibfried, S., Lewis, J., Obinger, H. and Pierson, C. (eds) *The Oxford Handbook of the Welfare State*. Oxford: Oxford University Press.

Bejan, I., Armasu, I. and Botonogu, F. (2014) *Tenant's Rights Brochure for Romania*. Report on behalf of the EU. Available at www.tenlaw.uni-bremen.de/Brochures/ RomaniaBrochure_09052014.pdf, accessed 8 August 2016.

BPIE (2011) *Europe's Buildings under the Microscope: A Country-by-Country Review of the Energy Performance in Buildings*. Brussels: Building Performance Institute Europe.

BPIE (2012) *How to Make Energy-Efficient Building Retrofit Happen in Romania?* Brussels: Building Performance Institute Europe.

BPIE (2014) *Renovating Romania: A Strategy for the Energy Renovation of Romania's Building Stock*. Brussels: Building Performance Institute Europe.

Castles, F. G., Leibfried, S., Lewis, J., Obinger, H. and Pierson, C. (eds) (2010) *The Oxford Handbook of the Welfare State*. Oxford: Oxford University Press.

CEB (2015) *Romania: Affordable Housing for Young People*. Available at www.coebank. org/en/news-and-publications/projects-focus/romania-affordable-housing-young-people/, accessed 12 July 2016.

Clapham, D., Hegedüs, J., Kintrea, K., Tosics, I. and Kay, H. (1996) *Housing Privatization in Eastern Europe*. Westport, CT: Greenwood Press.

Copenhagen Economics (2012) *Multiple Benefits of Investing in Energy Efficient Renovation of Buildings: Impact on Public Finances*. Brussels: Renovate Europe.

Cosima-Rughinis, A. (2004) *Social Housing and Roma Residents in Romania*. Budapest: Central European University/Centre for Policy Studies, Open Society Institute.

Czischke, D., Moloney, C., Turcu, C. and Scheffler, N. (2015) *Sustainable Regeneration in Urban Areas*. Paris: URBACT.

Dobrescu, E., Manea, G., Stefanescu, R. and Velter, V. (2011) 'Defining elements of sustainable development in the Romanian territory', *Review of General Management*, 14(2): 73–90.

EC (2006) *First National Action Plan for Energy Efficiency (2007–2010) – Romania*. Brussels: European Commission.

EC (2011) *Energy 2020: A Strategy for Competitive, Sustainable and Secure Energy*. Luxembourg: Publications Office of the European Union.

EC (2012) *Screening of Waste Management Performance of EU Member States*. Brussels: European Commission.

Energy Efficiency Watch (2013) *Energy Efficiency in Europe: Assessment of Energy Efficiency Action Plans and Policies in EU Member States: Country Report: Romania*. Brussels: Energy Efficiency Watch.

EU (2015). *Division of Powers: Romania*. Available at http://extranet.cor.europa.eu/ divisionpowers/countries/MembersNLP/Romania/Pages/default.aspx, accessed 12 July 2016.

Eurostat-SILC (2012) *Distribution of Population by Tenure Status, Type of Household and Income Group*. Luxembourg: Eurostat.

Eurostat (2013) *Sustainable Development in the European Union: 2013 Monitoring Report of the EU Sustainable Development Strategy*. Luxembourg: Eurostat.

Eurostat (2014) *House Price Index Database*. Available at http://appsso.eurostat.ec.europa. eu/nui/submitViewTableAction.do, accessed 12 July 2016.

Eurostat (2015) Real *GDP Per Capita, Growth Rate and Totals: Percentage Change on Previous Year, EUR per Inhabitant*. Available at http://ec.europa.eu/eurostat/tgm/ refreshTableAction.do?tab=table&plugin=1&pcode=tsdec100&language=en, accessed 12 July 2016.

Eurostat (2016a) *Eurostat Statistics Explained: Housing Statistics*. Available at http://ec.europa.eu/eurostat/statistics-explained/index.php/Housing_statistics, accessed 12 July 2016.

Eurostat (2016b) *Modal Split of Passenger Transport: % in Total Inland Passenger – km*. Available at http://ec.europa.eu/eurostat/tgm/refreshTableAction.do;jsessionid= 2ASR5Bx5iZNC9fZn_Z99Def4N1se78EuaFa1TJMPTjdfl4XzGxdl!-1493475499?tab =table&plugin=1&pcode=tsdtr210&language=en, accessed 12 July 2016.

Fatol, D. (2013) 'How Romania's five biggest cities are moving in the green direction', *This Big City*, 10 January. Available at http://thisbigcity.net/romanias-five-biggest-cities-moving-green-direction/, accessed 7 August 2016.

GovRo and UNDP (2008) *National Sustainable Development Strategy Romania 2012– 2020–2030*. Bucharest: Government of Romania and UNDP.

Grimshaw, D., Rubery, J. and Marino, S. (2012) *Public Sector Pay and Procurement in Europe during the Crisis: The Challenges Facing Local Government and the Prospects for Segmentation, Inequalities and Social Dialogue*. Manchester: University of Manchester.

Hills, J. (2007) *Ends and Means: The Future Roles of Social Housing in England*. London: London School of Economics.

Housing Europe (2015) *Social Housing in Europe: Romania*. Available at www. housingeurope.eu/resource-120/social-housing-in-europe, accessed 12 July 2016.

Howard, M. (2003) *The Weakness of Civil Society in Post-Communist Europe*. Cambridge: Cambridge University Press.

ICEMENERG and ANRE (2012) *Energy Efficiency Policies and Measures in ROMANIA. ODYSSEE- MURE 2010. Monitoring of EU and National Energy Efficiency Targets*. Bucharest: Energy Research and Modernizing Institute (ICEMENERG) and Romanian Energy Regulatory Authority (ANRE).

IEA (2010) *Passenger Car Ownership in the EEA*. Available at www.eea.europa.eu/data-and-maps/figures/passenger-car-ownership-in-the-eea, accessed 12 July 2016.

Kahrik, A. and Tammaru, T. (2010) 'Soviet prefabricated panel housing estates: areas of continued social mix or decline? The case of Tallinn', *Housing Studies*, 25(2): 201–219.

Klinckenberg Consultants (2010) *Making Money Work for Buildings: Financial and Fiscal Instruments for Energy Efficiency in Buildings*. Brussels: EuroACE.

MDRAP (2010) *Reabilitarea Termica a Blocurilor de Locuinte* [*Thermal Reabilitation of Multi-Family Blocks of Flats*]. Bucharest: Ministry of Regional Development and Public Administration.

MDRT (2012) *Cresterea performantei energetice a cladirilor. Bloc de locuinte. Standard de cost*. SCOST–04-01/MDRT. Bucharest: MDRT.

OECD (2013) *Trust in Government, Policy Effectiveness and the Governance Agenda.* Paris: OECD Publishing.

Rose, R. (2009) *Understanding Post-Communist Transformation: A Bottom up Approach.* London and New York: Routledge.

Rotariu, D. (2012) 'Reabilitarea Blocurilor de Locuinte in Romania'. Paper presented at Housing Retrofit: European Experience Exchanges, Maramures.

Szelenyi, I. (1993) *Urban Inequalities under State Socialism.* Oxford: Oxford University Press.

Tache, I. and Neesham, C. (2010) 'Romania and the Global Financial Crisis: impact and challenges', in Bartlett, W. and Monastiriotis, V. (eds) *South East Europe after the Economic Crisis: A New Dawn or Back to Business as Usual?* London: London School of Economics.

Tosics, I. and Hegedüs, J. (2003) 'Housing in south-eastern Europe', in Lowe, S. and Tsenkova, S. (eds) *Housing Change in East and Central Europe: Integration of Fragmentation?* Surrey: Ashgate.

Turcu, C. (2012a) 'Local experiences of urban sustainability: researching housing market renewal interventions in three English neighbourhoods', *Progress in Planning,* 78(3): 101–150.

Turcu, C. (2012b) 'Re-thinking sustainability indicators: local perspectives of urban sustainability', *Journal of Environmental Planning and Management,* 56(5): 695–719.

Turcu, C. (2015) 'Ar trebui să continuăm să investim fonduri publice în reabilitarea termică a blocurilor de locuințe în România? Cateva argumente in favoare' [Should we continue to direct public investment into the energy efficiency of apartment buildings in Romania? A few thoughts in favour], *OpenPoliticsRo.* Available at www.openpolitics. ro/platforma-de-discutii/reabilitarea-termica-blocurilor-de-locuinte.html, accessed 12 July 2016.

Turcu, C. (2016) 'Unequal spatial distributuion of retrofits in Bucharest's apartment buildings', *Journal of Building Research & Information.* DOI: 10.1080/09613218. 2016.1229894

Turcu, C. and Moloney, C. (2015) *Towards pro-environmental behaviour.* Working paper. Paris: URBACT.

Turcu, C. and Persson, A. (2015) 'Energy efficiency policy and action for multi-family residential building renovation in Central and Eastern Europe: the tale of four cities'. Paper presented at the ECEEE summer study, Toulon-Hyeres, 1–6 June.

Turcu, C. and Tosics, I. (2015) *Beyond Geographical Divides in Urban Europe.* Working paper. Paris: URBACT.

Turkington, R., van Kempen, R. and Wassenburg, F. (eds) (2004). *High-rise Housing in Europe: Current Trends and Future Prospects.* Delft: DUP Science.

Turnock, D. (1990) 'Housing policy in Romania', in Silliance, J. (ed.) *Housing Policies in Eastern Europe and the Soviet Union.* London: Routledge.

Ungureanu, I. (2014) 'Câte blocuri din Capitală sunt anvelopate şi câți bani au cheltuit primăriile pentru reabilitare termică de la începerea programului', *Adevarul,* 8 February. Available at http://adevarul.ro/news/bucuresti/reabilitare-termica-1_52f52cd5c7b855ff 5628656a/index.html, accessed 12 July 2016.

Valentin, A. and Spangenberg, J. (1999) 'Indicators for sustainable communities'.Paper presented at the Assessment Methodologies for Urban Infrastructure international workshop, Stockholm, 20–21 November.

van Kempen, R. (ed.) (2005) *Restructuring Large-Scale Housing Estates in Europe.* Bristol: Policy Press.

Vrabie, P. (2014) 'Primariile Capitalei vor reabilita de pana la sase ori mai putine blocuri. Cat de afectati vor fi producatorii din constructii si locatarii' *Wall Street*, 16 March. Available at www.wall-street.ro/articol/Social/163096/reabilitare-termica-2014.html, accessed 7 August 2016.

Zaman, G. and Georgescu, G. (2009) 'The impact of the global crisis on Romania's economic development', *Annales Universitatis Apulensis Series Oeconomica*, 11(2): 611–624.

10 Hungary

Iván Tosics

Introduction

Between the collapse of socialism in 1989/1990 and the mid-2000s, Hungary experienced a series of governments with very different positions on the political spectrum. In 2006, a left-wing government came to power, but in 2010 a right-wing party (Fidesz) achieved a landslide victory and two-thirds majority in parliament, a success that it replicated in 2014.

The economic changes from centrally planned to market economy started earlier in Hungary than in other countries in the region. Elements of market economy had been gradually introduced since 1980 (Tosics, 2005). At the beginning of the 1990s the transition to market economy was very quick, and in the 1990s Hungary acquired by far the most foreign direct investment compared with other post-socialist countries. Economic development slowed in the 2000s and Hungary could not maintain its initial advantage. In 2013, GDP/capita on PPP was 19,800 international dollars (IndexMundi, n.d.), behind Slovenia, the Czech Republic, Slovakia and Poland, and only exceeding Croatia, Romania and Bulgaria. Economic development was rapid from the early 1990s until 2008, at which point it stagnated. Today the country can be divided into three very different regions in terms of economic development: the Budapest region is on a par with Prague and Warsaw; the western part of the country resembles Slovakia; while the eastern part is comparable with Romania.

The economic effects of the financial crisis at the end of the 2000s were about average for an EU state (Hungary was less badly hit than Greece but more so than Poland). However, the effects were more substantial at sub-national level, where public expenditure decreased dramatically. It was reduced from 23 per cent of general government expenditure in 1995 to 14 per cent in 2013 (by contrast, the figures for the EU27 (no data for Croatia) were 31 and 32 per cent, respectively (EC, 2014: 143). Even so, sub-national government investments did not decline, probably due to the high share of EU financing of investments in Hungary. The country joined the EU in 2004, and over 70 per cent of public investments are financed by the organization (EC, 2014: 156).

The population at the end of the 2010s was 10 million, a decline from a high of 10.7 million in 1980. Forecasts predict further falls and a population of just

8.3 million by 2050 (UN, 2015). (Predictions for all post-socialist countries are significantly more negative than for many other countries.) These figures and projections reflect a combination of low fertility rates and high emigration rates. If the UN's prediction is correct, the demand for housing in Hungary is likely to decline.

The local government system is very fragmented in Hungary: there are over 3,300 municipalities. Despite their small average size, local governments have responsibility for all aspects of development (see, e.g., Tosics and Dukes, 2005). Although there is a directly elected middle tier of government (nineteen counties), their roles are very limited. Since the second half of the 1990s a new tier has been introduced (seven regions). Up till 2010, this level played an active role in planning and implementing EU programmes. However, since then, Viktor Orbán's right-wing government has practically eliminated the regions.

Hungary is a monocentric country with 17 per cent of the population (around 1,700,000 inhabitants) living in Budapest, the only large city (the secondary cities have populations of around 200,000). Regarding the welfare regime, it is not easy to categorize the country into any of the main models as there are elements of the liberal, social democratic and family-based South European regimes in the Hungarian system. As Hegedüs (2014: 263) states:

> welfare policy . . . did not originate from a consistent ideological model; policymakers did not follow a 'master plan'. Instead, welfare policies evolved in direct reaction to specific societal problems . . . This type of 'trial and error' or 'scrambling through' approach was more or less a general phenomenon in the region.

With respect to the social situation, Hungarian data show serious problems: regarding the proportion of the population living in severe material deprivation, Hungary has the third highest figure in the EU, behind only Bulgaria and Romania (EC, 2014: 72). It is important to note that people at risk of poverty and social exclusion live predominantly in rural areas: 28 per cent of rural and 20 per cent of urban populations fall into this category (the EU28 average is 10 per cent for both rural and urban populations).

There are only a few characteristics on which Hungary is not among the weakest among the eleven countries discussed in this volume. Although its Environmental Protection Index score (Table 1.1) is the second lowest, the value is not so far from the EU28 average. Meanwhile, its modal split score, showing the share of public transport among all mobility modes, is by far the highest of all eleven countries, reflecting one of the rare positive legacies of the socialist development model.

The housing system in Hungary

Comparative data used in this volume show a snapshot of the current situation in Hungary. This is very important in order to understand the relative position of the country within the EU. In the case of Hungary, however, the story would not be complete without briefly describing the dramatic changes of the last twenty-five years, during the transition from a socialist housing system into the present, free-market-oriented system. Having gained an understanding of these two very different housing systems, we can make comparisons between the models regarding their sustainability performance in housing and urban development.

The transition from the socialist to a market model of housing policy

In Hungary until the 1980s, the general rules and processes of a socialist housing model prevailed (Hegedüs and Tosics, 1996). Socialist urban development led to relatively compact cities with large proportions of new, prefabricated buildings in large, dense housing estates. Meanwhile, the old inner-city areas were neglected but survived. Within the housing system the share of state-owned housing was relatively large (one-third in Hungary, and two-thirds in Budapest). This large state rental sector was very inefficiently managed and run. Analysis at the end of the 1980s showed that the main housing problem in the transition countries was not a shortage of housing but the low quality, poor location and misallocation of the stock (Hegedüs and Tosics, 1996; Tosics, 2012).

In Hungary, the post-socialist transition in housing happened quite quickly, moving towards a market-dominated housing system in which public housing became very scarce and played only a residual role in the sense of Kemeny's (1995) definition.

In 1990, the share of state (public) rental housing was quite high in Hungary, and certainly higher than in most Western European countries. This fact, and the inefficiency of the state housing sector, made privatization unavoidable. As a result of large-scale privatization (Table 10.1), Hungary now has one of the smallest state rental sectors among all of the post-socialist countries.

Regarding the challenge to decrease the state housing sector, there were many options, as there are huge differences among the Western European capitalist countries in the composition of their housing stock according to tenure categories.

Table 10.2 shows three distinctive Western models: Northern social democratic; Central social market; and Southern private market (the three rows under 'Old EU countries'). Theoretically, post-socialist countries could have changed their tenure structure to any one of these models. Practically, though, most countries chose to move towards quick and large-scale privatization of the housing stock to the sitting tenants. Hence, post-socialist countries now most closely resemble the Southern model, with very small and marginalized public housing sectors (which tend to be too small to meet the housing needs of their poorest citizens).

The dramatic decrease of the share of public housing can be evaluated as follows:

> post-socialist societies in Central and Eastern Europe have opened up their housing systems to the (largely unregulated) market, while public housing as a means to manage social conflicts and decrease social inequalities has not gained sufficient impetus to become the driving force of a new housing regime.
>
> (Hegedüs *et al.*, 2015: 2)

Photo 10.1 The socialist housing system: large prefabricated housing estates and neglected inner-city areas (XVIII. Havanna housing estate in 2013 and IX. Middle Ferencváros in 2003).

Table 10.1 Housing tenure structure in Hungary, 1970–2011 (%)

	1970	1980	1990	2001	2011
Owner occupied	66.5	71.5	72.3	90.0	88.0
Municipality owned housing	33.3	28.3	19.0	3.7	2.7
Cooperative housing	–	–	3.7	1.0	1.0
Private rental and other	0.3	0.2	5.0	5.3	8.3
Total	100.0	100.0	100.0	100.0	100.0

Source: Hegedüs and Somogyi (2014), based on data from the Central Statistical Office, Hungary.

Table 10.2 Share of social (public) rental housing and the poverty rate in post-socialist countries, 2000s (%)

Countries	Social (public) rental housing	Poverty rate
Old EU countries		
NL, S, A	25–35	10–13
D, F, UK	15–25	14–18
ES, P, EL	1–5	19–23
Transition countries		
CZ, POL	10–12	15–25
H, EST	3–4	20–30
ALB, BUL, ROM	1–3	30–40

Sources: Author's estimates based on UN-ECE (2002, 2006).

As a consequence of this large-scale privatization, the public rental sector became too small to address social housing problems. Housing policy became largely 'indirect' in the post-socialist countries, as all types of housing problems (including the quality and sustainability of the housing stock and the maintenance of multi-family housing) can only be influenced with tools affecting the dominant privately owned housing stock.

Besides privatization, another indicator of the change in the housing system is the fluctuation of new housing construction. The extremely high output of the 1970s (reaching 100,000 units in 1975 at the height of panel construction) decreased to a very low level (no more than around 20,000 units per year) in the 1990s (Hegedüs and Somogyi, 2014). This was half of the 1 per cent of the total housing stock that was said to be needed to counteract normal amortization.

The years of economic boom between 2000 and 2008 led to a substantial increase in new construction due to almost unlimited bank loans (since the change in the political system in 1990, new housing construction has almost exclusively meant owner-occupied housing). Then came the collapse: the financial crisis led to unprecedentedly low figures of new output (below 10,000 units a year).

The housing finance system

Regarding the financing of new owner-occupied housing, banks started larger-scale mortgage lending activities in the second half of the 1990s. According to Hegedüs and Nagy (2014), three consecutive periods followed each other with very different national housing finance policies: the 2000–2004 subsidized mortgage market; the 2004–2008 FX (foreign exchange) mortgage market; and the 2008–2014 crisis management. The big (and, as later became obvious, unfortunate) change was the introduction of FX-denominated loans in 2004. In the case of Hungary, banks offered mainly Swiss franc-, but also some euro- and yen-denominated loans. Interest rates for these loans are based on the interest rates applicable to the currency in which the mortgage is denominated. The big risk with them is the unpredictability of the currency's exchange rate. As Figure 10.1 shows, the share of FX loans increased steeply between 2004 and 2008, until they represented almost 80 per cent of new lending in 2008.

There are many reasons why politicians ignored the risks of FX loans (for a detailed analysis, see Hegedüs (2014)). One of the main reasons for the extreme severity of the financial crisis in Hungary was the large stock of this type of loan – some €10 billion of foreign exchange loans were issued to Hungarian borrowers before 2008's GFC, and the share of non-performing loans increased from 5 per cent to 20 per cent between 2008 and 2014 (Hegedüs and Nagy, 2015: 1).

Social housing policies

From Table 10.2, it is easy to understand that affordability became one of the main housing problems, as the stock of social housing shrank dramatically to an insignificant level while the number of poor people increased (Photo 10.2). As a consequence, most poor people now live in the private housing sector as home owners or private tenants.

The most explicit social housing programme in the post-socialist period concerned housing allowances. The changes in this programme can be summarized (on the basis of Hegedüs (2014)) as follows. In 1993, as part of the Social Law, a local government-financed housing allowance scheme was introduced. However, this was not widely used, due to the absence of national co-financing. In 2004, the government introduced a new housing allowance programme, based on 90 per cent financing from the central budget. Another important programme was the energy consumption subsidy, which took into account that 70 per cent of housing costs are related to energy. This subsidy was integrated into the housing allowance scheme in 2012. Although these subsidy systems were increasingly means tested, the targeting was poor, so there was no guarantee that low-income households would be able to afford their housing costs after providing for their basic needs. Since March 2015, there has been no compulsory housing allowance system in Hungary: local government now decides to whom it gives support among families with housing and loan repayment problems. Kováts (2015) argues that this system will soon lead to significant problems in many poorer areas.

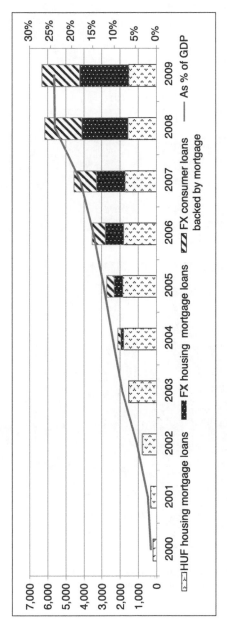

Figure 10.1 Mortgage developments in Hungary, 2000–2009.

Source: Hegedüs (2014).

Photo 10.2 The losers from the changes: poor people and low-quality housing
(VIII. Diószegi Street in 2009 and VIII. Hős utca in 2004).

Another, less significant project is the 'panel programme', introduced in 2004 with the aim of rehabilitating buildings built with prefabricated technology, mainly from an energy-efficiency perspective. As a result of this programme, 25 per cent of the stock of panel flats (190,000 housing units) was renovated to some degree by 2009 (Photo 10.3).

Photo 10.3 Energy-efficient renovation on housing estates in the 2000s (XVIII. Havanna housing estate in 2013).

The present housing system and its problems

Housing Europe (2015: 56) has provided an overview of the state of affairs in the Hungarian housing sector. The housing stock (12 per cent of which is vacant) is dominated by home ownership (92 per cent), while the real share of private renting may be around 8 per cent if non-reporting due to tax avoidance is included. The small municipally owned housing sector (3 per cent of stock) is concentrated in Budapest and the other larger cities. New housing construction dropped sharply during the financial crisis and reached a record low level of fewer than 10,000 units a year. The outstanding mortgage is only around 21 per cent of GDP (much lower than the EU average of 52 per cent); FX loans cover 72 per cent of the mortgage portfolio. Housing affordability is a large problem, with some 300,000 families in need of affordable housing, while almost two-thirds of people with an income below 60 per cent of the national average have arrears on their housing or mortgage repayments. Housing quality and comfort are still poor, despite improvements in the post-socialist period: Hungary has the second-highest rate of severe housing deprivation in the EU, after Romania. Finally, energy inefficiency in the housing stock is a problem.

From a spatial perspective the huge private housing stock and market-oriented housing policy display the usual problems of such systems: for instance, there is substantial differentiation in residential capacities regarding the renovation and modernization of the existing housing stock (condominium decision-making depends on the financial situation of the majority of owners, which does not allow for renovation in poorer urban areas). Socio-spatial segregation increased in Budapest between 2001 and 2010 (Tammaru *et al.*, 2015) as condominiums in better parts of the city became more homogeneous with higher-income families who welcome renovations, while the opposite is true in the poorer parts of the city (Photo 10.4).

Sustainable communities and housing in Hungary

Definitions of sustainability

Although many official documents address the topic of sustainability, in Hungary neither 'sustainable housing' nor 'sustainable communities' is well defined and the relevant policies have been very weak (or even non-existent in the first fifteen years of the transition).

In the 2000s, a comprehensive overview about the various aspects of sustainability was published (Bulla and Tamás, 2006). Besides environmental topics, social exclusion, intergenerational connections, the problems of youth and urban issues are also discussed. The overview includes an analysis of the sustainability aspects of the long-term development concepts of Budapest (Tosics, 2006), but housing is mentioned only indirectly, as an option for the sustainable development of the transitional belt of the city. No general definition of 'sustainable housing' or 'sustainable communities' is provided.

Photo 10.4 Increasing socio-spatial segregation in the 2000s: gentrified inner-city area versus poor area at the edge of the inner city (V. Október 6. utca in 2010 and VIII. Lujza utca in 2013).

In the second half of the 2000s, interest in environmental protection, climate change and sustainability gained momentum. In 2007 a new body – the National Council for Sustainable Development – was established. This organization developed the National Framework Strategy for Sustainable Development, which was adopted by the Hungarian parliament in 2013. The strategy defines 'sustainability' in quite narrow terms, mainly as a long-term, resource-management activity. In connection to the strategy, forty-four 'society indicators' are defined, of which only one is directly linked to housing (the share of housing units without bathrooms). The energy indicators do not include housing's energy consumption; only the transport sector is mentioned in this respect (KSH, 2015).

In 2012, a research overview on local sustainability strategies was published (Baják and Törcsvári, 2012). According to this document, in 2010 the National Council for Sustainable Development asked eight municipalities to prepare their own local sustainability agendas. Two years later, the initiatives of forty-eight municipalities and two micro-regions related to sustainable development were accessible on the internet. In these strategies there are eight topics of major importance:

> These are energy management and conservation, transportation, the protection of the built environment, natural values, the protection of municipal green spaces, waste management, water management, the development of environmental awareness among citizens and air quality ... Beside the aspects that are related to environmental sustainability, economic (for example the development of the economy, tourism) and social (employment, education, health care) objectives also appear in the strategies which define themselves as local agendas.
>
> (Baják and Törcsvári, 2012: 6)

Housing is not mentioned directly. The protection of the built environment is a general aim while land policy is rarely mentioned in local documents.

In 2012, in the course of the national planning process for the 2014–2020 EU programming period, a lobby document (Hungarian Housing Lobby, 2012) was published by twelve important players in the construction and housing industry (Associatiation of Home Builders, Assoication of Building Material Companies, Association of Real Estate Companies, Trade Union of Builders, Hungarian Energy Efficiency Office, Chamber of Architects and so on). The document emphasizes the importance of eligibility of housing-related interventions for EU financing in the 2014–2020 Cohesion Policy regulation. It mentions as desirable interventions:

- the energy-efficient modernization and renewal of residential and public buildings;
- the construction of publicly owned rental housing and public buildings; and
- targeted interventions to improve deprived areas with urban renewal projects for residential and public buildings.

The publication of this document shows the absence/weakness of national policy determination towards interventions to improve housing sustainability.

Hungarian housing and urban policies from a sustainability perspective

National-level policies towards sustainability have been very weak and narrow in scope in Hungary, regarding territorial, urban and housing policies as well. The transition in the 1990s, dominated by decentralization and privatization, constrained the possibilities of sustainable development:

> in the process of decentralization, local governments attained real independence ... but the main consequence of the decrease of direct central control was fragmentation (the decisions of the settlements could not be coordinated, the public interests connected to the level of the larger areas were not represented satisfactorily).
>
> (Tosics, 2004: 75)

Territorial/regional development policies were unsuccessful in reducing the territorial development differences within the country. Due to the spatial patterns of FDI (foreign direct investment), the dominant position of Budapest within the urban structure has increased. Similarly, the very large territorial differentiation within the country, the 'west–east slope', has grown, insofar as the eastern part of the country has become even poorer, compared to the western part (Ekés *et al.*, 2013). Although significant funds have been available from the EU's Cohesion Policy to support the lagging parts of Hungary since 2004, there is a growing number of 'forgotten settlements', predominantly Gypsy ghettos where all of the residents are unemployed.

Regarding *urban development*, the sustainability aspect has been very weak over recent decades and the recent national sustainability policy has practically no urban angle at all. Since the administrative decentralization of 1990, settlements have made their own decisions about rezoning areas within their administrative borders. There are many analyses (e.g. Kovács and Tosics, 2014) about the consequences of this extreme application of the subsidiarity principle. Until the late 2000s suburbanization processes accelerated, not only regarding residential mobility but also concerning office and commercial functions. Hungary (similarly to other post-socialist countries) has been unable to enforce sustainability principles for the coordination between new urban development and the existing infrastructure as investors and developers usually reject any preconditions for their developments, which leads to the dominance of green-field over brown-field projects.

In the course of the 2000s there was an attempt to tackle the lack of coordination between planning and development among neighbouring settlements in the very fragmented Hungarian settlement structure. A system of micro-regions was gradually developed (with 173 micro-regions covering the 3,300 Hungarian settlements), with additional state subsidies given to those where all the settlements agreed to cooperate in running the public services and preparing development plans. This system functioned only on a voluntary basis as Fidesz (which was in opposition at the time) refused to join with the government to make it compulsory. Despite encouraging results – almost all of the micro-regions established cooperation mechanisms – after 2010 the new Fidesz government eliminated the system, replacing it with a similar territorial level with a different name but only administrative functions. This unfortunate change has had negative consequences around Budapest: Pest County now has to deal with 187 settlements on a one-to-one basis. Obviously, this is much more difficult than dealing with the previous sixteen micro-regions, each of which had two to three experts in public services and urban development issues.

Transport policies have had a large impact on urban development, and they can be considered as another 'sustainability loss'. One of the principal legacies of socialism is Hungary's extensive public transport system, largely built at the beginning of the twentieth century. During the socialist period this system was preserved (although not modernized) and extended with a few metro lines in Budapest. However, the market-based period led quickly to the dominance of car-oriented development (Tosics, 2007). As a consequence of the low-quality public transport and insufficient modernization, tram lines were shut down one after the other. Hungary's larger cities could replace their obsolete trams only with second-hand carriages bought cheaply from other countries (mainly the former East Germany). The deteriorating tracks could not be maintained in a similar way, however. Only over the last few years has substantial money (almost exclusively provided by the EU's Cohesion Policy) been spent on modernization and some extensions to the public transport network. Hungary's cities still have a large proportion of public transport (good modal split), but the opportunity to pursue a more compact, public transport-based development has now disappeared.

This short overview of national sustainability-related policies has shown that the post-socialist period started with neoliberal, laissez-faire approaches in all policy areas, minimizing the influence of national government over market processes in urban and housing development. However, in the course of the 2000s some interesting policy initiatives emerged in Hungary, including the national system of housing subsidies and the system of micro-regions in order to initiate territorial cooperation. These policy innovations had important advantages from a sustainability perspective and could be considered good practice on a pan-European level. Unfortunately, though, most of them did not prevail for very long: they did not become sufficiently embedded within policy thinking and practice. Such multi-governance structures need time to develop fully, but they were swept away as national and local politics moved towards top-down, power-based, authoritarian government at the beginning of the 2010s.

Interpreting Hungarian sustainable housing and communities
indicators

The comparative data in Table 1.1 present information about the eleven countries that feature in this volume from a variety of perspectives. A quick overview of the data leads to the following impressions about Hungary in comparison to the other ten countries:

- In GDP per capita terms, Hungary is the second-poorest country after Romania, not even reaching half of the EU28 average. However, this differ-ence would be slightly less if calculating GDP on the basis of Purchasing Power Parity.
- The unemployment figure is moderate, and less than the EU average. However, this relatively good indicator value has been achieved through the present government's huge efforts to force unemployed people to participate in public work – a system that does not bring the unemployed any closer to real jobs.
- The housing system is rightly labelled 'super home ownership', with over 90 per cent of the population living in owned homes.
- The welfare regime is in transition and is a mixture of the various models.
- The level of poverty and social exclusion is the second highest among the eleven countries investigated, and well above the EU28 average.
- The modal split of passenger transport is the only indicator in which Hungary ranks highest among the investigated countries. It has by far the highest share of public (train, bus) transport as opposed to individual car use.
- In the Environmental Protection Index Hungary is again the second-worst country after Romania, although not far from the EU28 average.
- According to the housing cost overburden rate, the ratio of housing costs to income is higher than average in Hungary, especially for tenants and those who have mortgages.
- Regarding housing quality, Hungary has the second-lowest value (again after Romania), indicating huge problems with the physical quality of and overcrowding within the country's dwellings.

Other information sources allow for additional comparisons between Hungary and other EU countries. According to Housing Europe (2015), the construction price level in Hungary is 44 per cent of the EU28 average, similar to the relative level of GDP (40 per cent of the EU28 average). There are several aspects in which Hungary is among the worst/weakest of all EU countries:

- regarding arrears on mortgages among those below 60 per cent of median income, Hungary has the second-highest score (63 per cent), after Greece, while the EU28 average is 32 per cent;
- regarding severe housing deprivation among those below 60 per cent of median income, Hungary again has the second-highest score (45 per cent), after Romania, while EU28 average is 13 per cent; and

- the energy performance of the country's housing stock is so poor that Hungary (along with the Baltic states) has received recommendations and advice on the subject from the European Commission (hence, fuel poverty is also a significant issue).

Barriers, drivers and challenges for sustainable communities in Hungary

One of the main barriers against sustainable development was the *quick, large-scale privatization of public housing*. This, combined with a lack of appropriate national policies, led to negative consequences in housing and urban development. This can be illustrated with the example of the large housing estates, which correspond to around a fifth of Hungary's total housing stock, and a third of the stock in Budapest. These estates have some good sustainability potential because of their environmentally friendly district heating system, their large proportion of green areas and their good public transport connections. After the privatization of the housing stock, however, each panel building became an individual condominium and interest in the common areas of the estates declined. Some buildings have left the (not very efficient and therefore quite expensive) district heating system, and the green areas have been increasingly occupied by parked cars. National subsidies have been introduced for the energy-efficient regeneration of the panel buildings and local governments have launched programmes (largely funded by EU money) to improve the public areas within large housing estates only over the last decade.

Another important barrier against sustainable development was the *decentralization of all decision-making rights to the lowest level – the very fragmented system of local governments*. This led to the emergence of urban sprawl in the Budapest area (Photo 10.5). Until 2005, all local municipalities around the capital had full discretion about their territory and could rezone agricultural land into residential or commercial use according to their own wishes. In 2005, after many years of debate, a national law was finally passed which constrained these rezoning rights insofar as rezoning in the area of the Budapest Agglomeration now needed parliamentary approval. However, by then it was too late. During the protracted discussions about the new law, all of the settlements around Budapest rezoned the territories they felt they would need for residential and commercial developments in the future. According to data from Péter Schuchmann (around 2010, in an unpublished presentation), the reserved area for future residential development amounts to 6,740 hectares (10 per cent of all residential areas) and the reserved area for future commercial development amounts to 7,450 hectares (one-third of all commercial areas). This shows that, although in principle the 2005 law set strict limits on further urban sprawl, it will not be effective for at least another fifty years, due to the large reserves the suburban settlements had already rezoned.

The dangers of urban sprawl were well known in Hungary at the beginning of the 1990s, when the post-socialist period began. Over the next fifteen years, however, a neoliberal, laissez-faire ideology prevailed at the national level and

Photo 10.5 Urban sprawl around Budapest (Törökbálint, Tükörhegy, in 2008 and aerial photo in 2013).

tiny local governments were given total freedom to decide how to proceed with urban development. The immense sustainability problems of today could have been avoided with a more visionary public policy that subordinated housing policy tools (mortgages, subsidies) to long-term urban development goals and oriented new housing construction to already developed areas with existing infrastructure and public transport connections in the functional urban area of Budapest. Furthermore, imposing a tax on real-estate value increase would have helped to address the issue that the advantages and externalities (disadvantages) of development affect settlements in different ways.

There was very little likelihood of Hungary's government introducing such visionary policies, however. The only tool Budapest possesses to influence urban development is the Strategic Development Concept (first approved in 2003, and adopted in a new version in 2014). However, the territorial scope of this concept is limited to the administrative boundaries of Budapest, so it covers only the 1.7 million people in the city centre, not the almost 1 million who live within the functional urban area of the city. Moreover, in reality, urban development is largely determined by private developers while the municipality can only influence the use of funds from the EU's Cohesion Policy (which are predominantly spent on infrastructure development).

The figures relating to the state of affairs in Hungarian housing illustrate the main problems of the 'super home ownership' housing structure: in the absence of any sizeable public (affordable) rental stock, a very large proportion of the population – the poor home owners – have difficulty covering their housing costs, even though the Hungarian housing stock is of very low quality in comparison with other EU countries. Thus, Hungary's main problems are the quality and state of deprivation of the existing housing stock and the low level of affordability, rather than the quantity of dwellings.

The affordability problem was further aggravated by the problem of the FX loans (accounting for 72 per cent of the mortgage portfolio), especially in the course of the GFC, which led to a worsening in the exchange rate of the forint against the Swiss franc. The Hungarian government compulsorily converted all FX loans into HUF loans in November 2014 (just before a huge further increase in the value of the Swiss franc in January 2015), putting the largest burden on the banks. However, after the 2014 elections the government's housing programme failed to meet other housing policy targets. The social safety net became even weaker, especially after the 2015 withdrawal of the general housing allowance system. Now the allowance covers only a fraction of people, based on decisions made by the local municipalities. In 2016, the government announced a new support programme for improving housing conditions of families with at least three children. First assessments indicate that this programme might give an impetus to the housing market (new construction and enlargement of existing housing), but it will not affect affordibility problems as the conditions favour mostly middle-class families.

The analysis in this chapter has shown that in the case of Hungary a more sustainable housing system will be achieved only through a substantial increase in the number and proportion of affordable housing units. This can best be achieved through the enlargement of the (below-market-rent) stock of rental units. Furthermore, this does not require significant new construction; it could be done mainly on the basis of the existing housing stock, in which some half a million units lie vacant at present (Housing Europe, 2015).

Housing Europe's (2015) policy-development overview cites the introduction of new rental housing initiatives as an important aspect of new housing policy in the EU, with reference made to several countries in Eastern and Southern Europe (where rental stock is very small). However, Hungary is not mentioned among the countries that plan to introduce new, socially oriented rental housing policies. (Such a policy is far from the philosophy of the present right-wing government, which has made steps in the opposite direction.) On the other hand, there are informal and bottom-up efforts in Hungary to reuse parts of the empty owner-occupied housing sector through innovative new rental housing agencies (Hegedüs *et al.*, 2015).

Social conditionality towards integrated development: local anti-segregation planning in Hungary in the 2000s

This section tells the story of a unique 'social proofing' method that was applied in the second half of the 2000s, raising social (anti-segregation) conditions for cities in order to access EU funding. In earlier parts of this chapter, it has been shown that there are no explicit sustainability strategies for urban development and the housing sector in Hungary. This does not mean, however, a total absence of such considerations in the country. In Hungary, 'integrated development' is the closest term and policy to the definition of sustainability.

In the 2000s, after the first decade of transition, the problems of the dominant market mechanisms were apparent. Around the time of joining the EU (2004), it was recognized that integrated planning approaches were required to create more balance between the economic, environmental and social aspects of development. The EU accession created a good opportunity to put this idea into practice: local municipalities were eager to access Cohesion Policy/Structural Fund resources and were prepared to accept some conditions in return. Below, I describe how local leaderships in the second half of the 2000s started to move towards more social sustainability in order to access these EU funds. (The description is based on Tosics (2011). See also EC (2011).)

In Hungary in 2007, the preparation of an Integrated Urban Development Strategy (IUDS) was made compulsory for cities applying for EU Structural Funds for urban renewal actions as part of the Regional Operational Programme 2007–2013. The IUDS was a medium-term (seven–eight-year) strategic implementation-oriented document with sectoral and territorial aims. It had to be discussed and approved by the municipal assembly to ensure legitimacy.

One of the novelties of the IUDS was that cities had to prepare anti-segregation plans. Such plans had to contain the delimitation of segregated areas and areas threatened by deterioration and segregation on the basis of indicators devised by the national administration. They also had to include assessments of the delimited areas and of the potential social impacts on these areas of the envisaged developments and policies. Moreover, anti-segregation programmes had to be prepared, including a vision for a regeneration or elimination of the degraded areas and for the main directions of interventions. A complex set of tools had to be used with a focus on local housing, education, social and healthcare conditions.

In order to contribute to the required planning tasks of local governments the National Ministry for Development and the Economy (NFGM) established a group of anti-segregation experts and commissioned the preparation of an *Anti-Segregation Guidebook*. The activities of the experts were financed by the ministry, which allowed them to help the cities to draft their plans. The anti-segregation experts also had the task of guaranteeing the quality of these plans: without their counter-signature, the cities could not receive the requested EU funding for their urban renewal actions. In this way the social (anti-segregation) dimension became a condition for accessing EU funding. (This innovative means of applying strong social conditionality was unknown in other EU countries.) The new legislation generated strong opposition among local politicians, mainly due to the rejection of the idea to deal with the most segregated areas.

The Hungarian IUDS can be considered a success: over 200 cities prepared integrated development strategies, including anti-segregation plans. The latter became key elements of the general 'equal opportunities policy', with a recognition that cities could be forced to think about how to reduce segregation (in the broader sense, how to support social sustainability) only if this was posed as a general condition for accessing EU funding.

Such innovative approaches and tools have to be based on strong and high-level political commitment, and they are invariably threatened by political change. This was especially true in Hungary, where 2010 marked a fundamental political shift to the right. Although the system of local integrated planning still exists, in the new programming period the 'social proofing' through anti-segregation planning is less important as its content is not as tightly regulated as before. Moreover, the anti-segregation experts have been removed from the system.

Conclusions

Post-socialist countries offer a rare opportunity to compare the 'sustainability performance' of two very different housing systems: the socialist model (before 1990) and the market-based model of the last twenty-five years. From an *economic* perspective, the socialist model was much less efficient than the market model, as new public housing in the socialist period was built with large state subsidies in areas and forms that consumers did not really want. In public housing investment decisions, land values were totally ignored and inner-city areas with potentially large market and heritage values were allowed to deteriorate. Also, the management of public housing was very inefficient: systematic renovation was replaced by politically decided prestige improvements. While these inefficiencies disappeared in the market model, unfortunately so did the public housing sector, which was largely privatized.

Regarding the *environmental* dimension, the opposite seems to be the case: the socialist model was in some respects more sustainable than the market model. In the socialist period, cities were developed in a compact way: a large proportion of new housing in cities comprised large housing estates which had high densities, were heated with environmentally friendly district heating systems, had substantial green areas and were well served by public transport, making individual car use unnecessary for travel to work. However, environmental sustainability in the socialist period was, of course, achieved on a much lower level than is required today. Due to cheap energy prices, the insulation of buildings was very poor, and the green areas and public transport systems were developed/ running at a relatively low standard. Under the market model, the development of environmental infrastructure has accelerated, but mostly in suburban areas without any consideration for compact development.

Considering *social* sustainability, comparison does not lead to clear superiority of either of the two housing models. Housing affordability was in general less of a problem under the socialist model: poor families could afford state rental housing as rents were kept artificially low. Waiting lists gave poor families, especially those with a number of children, at least some chance of obtaining a state rental flat. At the same time, however, a large strata of society was totally excluded from this option and had to solve their housing needs by self-building or looking towards the subordinated private sector. Socio-spatial segregation was less of a problem during the socialist model, especially in the earlier period, when new, more comfortable state housing was built in less prestigious outer areas, while formerly high-status inner areas deteriorated and became more socially mixed. In the last two decades of the socialist period, however, as the market became more important, the segregation of higher-status groups started to increase. The current market model, with its dominant owner-occupied housing sector, offers gradual opportunities to enter the system as the housing stock and prices are very differentiated, with relatively low prices at the bottom end of the housing hierarchy (deteriorated inner-city housing, peripheral housing estates). On the other hand, the segregation of the middle class and richer families has increased, as they are now concentrated in the best, fully renovated parts of the inner city, new gated garden estates or high-status suburban areas. Furthermore, there is a growing tension between the hundreds of thousands of vacant flats and the hundreds of thousands of families without suitable housing.

This comparison has shown that the last quarter of a century, since the collapse of the socialist model, has brought contradictory changes regarding the sustainability of the Hungarian housing system. Although housing standards and quality of life have increased for many, social and territorial differentiation has also risen. Housing policy has not been prioritized over the past twenty-five years, as there has been an unstated assumption that the 'invisible hand' of the market will solve any problems. The resultant lack of political will to regulate the market in the public interest, including environmental and social sustainability, has led to the numerous problems and contradictions that have been discussed in this chapter.

A closer look at Budapest reveals that free-market conditions, combined with decentralization of land-use decisions to the level of tiny municipalities, have led to urban sprawl. Limited public control, a lack of sustainability in urban planning – and especially in housing policy – combined with no regulations to constrain the free market have resulted in the public interest being subordinated to the interests of suburban settlements and individual households.

Some analysts of the transition period (including the author of this chapter) hoped that after the establishment of the market-based urban and housing systems in 1990, much-needed public control over urban development and a functioning housing market would be introduced over time. It was hoped that this would also result in the introduction of more specific sustainability considerations. However, successive national governments had different aims and almost never continued the programmes of their predecessors. In the second half of the 2000s, there were some attempts at the national level to address the most pressing sustainability problems (e.g. launching a national system of housing subsidies, introducing micro-regions to foster territorial cooperation within functional urban areas, and developing social proofing of development plans to ensure more justice in urban development). Unfortunately, though, these initiatives failed to develop into well-established policies. Then, in 2010, an anti-liberal political party won the general election and secured a two-thirds majority in parliament. This concentration of political power was not used to establish policies for more environmental and social sustainability. Rather, the government's populist and pragmatic (to maintain power) considerations have led to further increases in social inequality and contradictory regulations regarding the environment. Public control over market processes has increased, but, unfortunately, not for the sake of more environmental and social sustainability.

By the mid-2010s, there were ever more examples of radical changes in the leaderships of Europe's large cities. The new mayors of Prague, Madrid and Barcelona, as well as the representatives of the Urban Movement in most Polish cities, have totally new views about urban development, placing social and environmental sustainability at the top of their agendas. Such new visions, especially if coupled with similar changes in national policy-making, might well lead to more sustainable urban and housing development. At present, however, there are no signs of such a change taking place in Hungary.

References

Baják, I. and Törcsvári, Zs. (2012) 'Local sustainable development programs in Hungary', *Periodica Oeconomica*, 2012: 39–49. Available at: www.gti.ektf.hu/anyagok/po/2012/PO2012_Bajak_Torcsvari.pdf, accessed 12 July 2016.

Bulla, M. and Tamás, P. (eds) (2006) *Fenntartható Fejlődés Magyarországon* [*Sustainable Development in Hungary*]. Budapest: Új Mandátum Könyvkiadó.

EC (2011) *Cities of Tomorrow: Challenges, Visions, Ways Forward*. Brussels: European Commission.

EC (2014) *Investment for Jobs and Growth: Sixth Report on Economic, Social and Territorial Cohesion*. Brussels: European Commission.

Ekés, A., Gerőházi, É., Gertheis, A., Hegedüs, J., Tosics, I. and Tönkő, A. (2013) 'The state of the Western subregion's cities', in Maseland, J., Kayani, L., Rochell, K., Adamski, J. and Janas, K. (eds) *The State of European Cities in Transition 2013: Taking Stock after 20 Years of Reform*. Nairobi: UN Habitat, pp. 52–99.

Gerőházi, É., Hegedüs, J., Szemző, H., Tomay, K. and Tosics, I. (2012) 'The impact of European demographic trends on regional and urban development', in Martinez-Fernandez, C., Kubo, N., Noya, A. and Weyman, T. (eds) *Demographic Change and Local Development: Shrinkage, Regeneration and Social Dynamics*. Paris: OECD Local Economic and Employment Development Working Paper Series.

Hegedüs, J. (2013) 'Social housing in Hungary: ideas and plans without political will', in Hegedüs, J., Lux, M. and Teller, N. (eds) *Social Housing in in Transition Countries*. London: Routledge, pp. 180–194.

Hegedüs, J. (2014) '"Unorthodox" housing policy in Hungary – is there a way back to public housing?', in Jie Chen, J., Stephens, M. and Yanyun Man, J. (eds) *The Future of Public Housing: Trends in the East and the West*. Germany: Springer-Verlag, pp. 259–278.

Hegedüs, J. and Nagy, G. (2014) 'The effect of GFC on tenure choice in a post-socialist country: the case of Hungary'. Paper presented at Urban Studies Foundation seminar on 'The Edges of Home Ownership', Delft, 21 December.

Hegedüs, J., Horváth, V. and Somogyi, E. (2015) 'Social rental agencies – a possible breakthrough in social housing provision in post-socialist countries: the case of Hungary', *European Journal of Homelessness*, 8(2): 41–67.

Hegedüs, J. and Somogyi, E. (2014) 'Moving from an authoritarian state system to an authoritarian market system: housing finance milestones in Hungary between 1979 and 2014', in Lunde, J. and Whitehead, C. (eds) *Milestones in European Housing Finance*. Chichester: John Wiley & Sons, pp. 201–218.

Hegedüs, J. and Tosics, I. (1996) 'Disintegration of the East-European housing model', in Clapham, D., Hegedüs, J., Kintrea, K. and Tosics, I. with Kay, H. (eds) *Housing Privatization in Eastern Europe*. Westport, CT: Greenwood Press, pp. 15–40.

Housing Europe (2015) The *State of Housing in the EU 2015*. Brussels: Housing Europe.

Hungarian Housing Lobby (2012) 'EU-források bevonása az energiahatékony és fenntartható építés számára: építőipari élénkítés érdemi költségvetési források nélkül a 2014–2020 közötti ciklusban' [The inclusion of EU resources into energy efficient and sustainable construction: initiating the building industry without using budget resources in the 2014–2020 period]. Available at www.lakasepitesert.hu/upload/EU_forrasok_energiahatekony_epites_2012_11_06.pdf, accessed 12 July 2016.

IndexMundi (n.d.) *Hungary GDP – per capita (PPP)*. Available at www.indexmundi.com/hungary/gdp_per_capita_%28ppp%29.html, accessed 12 July 2016.

Kemeny, J. (1995) *From Public Housing to the Social Market*. London: Routledge.

Kovács, Z. and Tosics, I. (2014) 'Urban sprawl on the Danube: the impacts of suburbanization in Budapest', in Stanilov, K. and Sykora, L. (eds) *Confronting Suburbanisation, Urban Decentralization in Post-Socialist Central and Eastern Europe*. Chichester: Wiley Blackwell, pp. 33–64.

Kováts, B. (2015) *Az önkormányzati hatáskörbe került lakhatási támogatások vizsgálata 31 önkormányzat példáján* [*Survey of the Locally Allocated Housing Allowances in 31 Municipalities*]. Magyarország: Habitat for Humanity.

KSH (2015) *A fenntartható fejlődés indikátorai Magyarországon, 2014* [*The Indicators of Sustainable Development in Hungary, 2014*]. Központi Statisztikai Hivatal. Available at www.ksh.hu/docs/hun/xftp/idoszaki/fenntartfejl/fenntartfejl14.pdf, accessed 12 July 2016.

Tammaru, T., Marcinczak, S., van Ham, M. and Musterd, S. (eds) (2015) *Socio-Economic Segregation in European Capital Cities: East Meets West*. London: Routledge.

Tosics, I. (2004) 'European urban development: sustainability and the role of housing', *Journal of Housing and the Built Environment*, 19: 67–90.

Tosics, I. (2005) 'City development in Central and Eastern Europe since 1990: the impact of internal forces', in Hamilton, H., Dimitrovska, Andrews, K. and Pichler-Milanović, N. (eds) *Transformation of Cities in Central and Eastern Europe: Towards Globalization*. New York, Tokyo and Paris: United Nations University Press, pp. 44–78.

Tosics, I. (2006) 'A nagyvárosok fenntartható fejlődése és Budapest jövőképe' in Bulla, M. and Tamás, P. (eds) *Fenntartható fejlődés Magyarországon. Stratégiai kutatások – Magyarország 2015*. Budapest: Új Mandátum Könyvkiadó, pp. 420–443.

Tosics, I. (2007) 'City-regions in Europe: the potentials and the realities', *Town Planning Review*, 78(6): 775–796.

Tosics, I. (2011) 'Governance challenges and models for the cities of tomorrow'. Paper prepared for the European Commission – DG Regional Policy for the Cities of Tomorrow Programme, January.

Tosics, I. (2012) 'Housing and the state in the Soviet Union and Eastern Europe', in Smith, S., Elsinga, E., Fox O'Mahony, L., Eng, O., Wachter, S. and Hamnett, C. (eds) *International Encyclopedia of Housing and Home*, Vol. 3. Oxford: Elsevier, pp. 355–362.

Tosics, I. (2015) 'Housing renewal in Hungary: from socialist non-renovation through individual market actions to area-based public intervention', in Turkington, R. and Watson, C. (eds) *Renewing Europe's Housing*. Bristol: Policy Press, pp.161–186.

Tosics, I. and Dukes, T. (2005) 'Urban development programmes in the context of public administration and urban policy', *Tijdschrift voor Economische en Sociale Geografie*, 96(4): 390–408.

UN (2015) *World Population Prospects, the 2015 Revision*. New York: UN Department of Economic and Social Affairs, Population Division. Available at http://esa.un.org/unpd/wpp/Publications/Files/Key_Findings_WPP_2015.pdf, accessed 12 July 2016.

UN-ECE (2002) *Bulletin of Housing and Building Statistics for Europe and North America 2002*. Geneva: United Nations Economic Commission for Europe.

UN-ECE (2006) *Bulletin of Housing and Building Statistics for Europe and North America 2006*. Geneva: United Nations Economic Commission for Europe.

11 Switzerland

Margrit Hugentobler

Introduction

This chapter is based on the premise that progress toward sustainable urban communities is largely dependent on a country's political system as well as its economic wellbeing. The former determines the kind of policies pursued in order to achieve environmental, social and economic sustainability goals and the measures taken. The latter influences the amount and types of resources invested in trying to reach these goals. This chapter highlights how the rather unique Swiss democratic decision-making system influences policies toward sustainable urban development at the national, cantonal and local levels. The particular focus is on local policies as they relate to the provision of sustainable housing in a more encompassing context of sustainable urban neighbourhood and city development.

With limited natural resources apart from beautiful landscapes and water from lakes and rivers that are relevant for electricity generation, Swiss economic development has historically been based on value-added products, quality and innovation. Examples are the watch industry, the manufacturing of precision tools/instruments, the pharmaceutical industry, as well as the growth of the high-tech sector as part of the digital revolution. Important service-related sectors are finance and tourism. In the year 2012, 72.5 per cent of employment was in the service sector, 20.3 per cent in industry, and only 3.5 per cent in agriculture and forestry (Statista, 2015).

The Swiss economy grew continually from the mid-1990s until the US subprime crisis in 2008. The following period of economic decline lasted less than two years and affected mostly larger Swiss banks due to their investments in high-risk financial market instruments abroad. Between 2010 and 2014, the economy prospered again, with an average annual GDP growth rate of around 2 per cent (Swiss Statistics, 2016a). According to Eurostat statistics, in 2013 the GDP per capita income in Switzerland was €63,800, ranking second only to Norway. The 3.2 per cent unemployment rate in 2014 (Swiss Statistics, 2015a) was the lowest of all European countries (see Table 1.1). An indicator of economic growth, unemployment in Switzerland averaged 3.33 per cent between 1995 and 2015 (Trading Economics, 2015). At the same time, the gap between the highest and the lowest incomes continuously grew, with the top 20 per cent earning more than four times as much as the bottom 20 per cent (OECD Better Life Index, 2015). A report on poverty in Switzerland showed that the disposable monthly income of 7.8 per cent of all households was below the poverty line in 2010, affecting over half a million of people. Comparative European Union statistics define the poverty line at slightly below €2,000 disposable income per month for a single person in Switzerland. Households most at risk are single-parent families, persons with low educational levels, and single-person households below sixty-five years of age (Swiss Statistics, 2012). Even more skewed than the income distribution is the distribution of wealth. In this regard Switzerland is one of the most unequal developed countries. According to the *Global Wealth Report*, in 2015 the richest 10 per cent of the population owned 71.6 per cent of the accumulated wealth (Credit Suisse Research, 2015).

Switzerland has a democratic, decentralised, federal political system. The legislative body comprises a multi-party system in which five major parties represent the majority of the voting population. The seven-member executive government represents the four biggest parties. In contrast to other European countries, where party coalitions form majority governments, government decisions in Switzerland reflect issue-related shifting alliances between the parties. Switzerland has been classified by a number of authors as having a liberal welfare regime (Arts and Gelissen, 2010). However, the Swiss system of social protection has been substantially modified since the early 1970s. Major programmes have been altered, some even fundamentally, resulting in a shift from the liberal welfare state regime to one that has close affinities to a conservative model (Bogedan *et al.*, 2010). Nevertheless,

> diversity amongst the approaches exists, and while some cantons follow a liberal minimalist approach, others are more generous and egalitarian in their orientation, and still others have a clearly conservative orientation based on a traditional vision of the family. Hence some of the variation of Western European welfare states could be reproduced within Switzerland itself.
>
> (Armingeon *et al.*, 2004: 22)

Switzerland had a population of 8.08 million in 2013 (Swiss Statistics, 2015c). It continues to grow, largely due to immigration and economic prosperity and facilitated by Switzerland signing the Schengen Agreement in December 2008. Migration has always played an important role in Switzerland. Reaching 24.3 per cent in 2014, Switzerland has one of the highest proportions of immigrants in Europe. This is partially explained by the fact that it includes second-generation persons born in Switzerland who have not yet applied for Swiss citizenship. The origins and characteristics of immigrant groups reflect Swiss economic growth patterns, as well as political and economic crises in other countries. Persons immigrating between the 1960s and the turn of the century largely came from Southern European and Balkan countries. From 2000 onwards, patterns changed, with highly qualified immigrants from Western Europe, mostly Germany, seeking jobs in the service sector (Swiss Statistics, 2015b).

The housing system in Switzerland

In 2013, the Swiss housing stock consisted of 57.5 per cent single-family and semi-detached dwellings, 26 per cent multi-unit dwellings, with the remaining 16.5 per cent mixed-use buildings (Swiss Statistics, 2014). Much of the housing stock is relatively old. Close to 80 per cent of the buildings in Switzerland were constructed before 1990. This percentage corresponds to both the European and the German average (Building Performance Institute Europe, 2011). Almost half (49 per cent) of the Swiss residential building stock dates from before 1960. It is worth noting that around one-third of the buildings dating from before 1920 have not yet been renovated. Renovation rates vary greatly in different cantons. While in Tessin, in southern Switzerland, 71 per cent of single-family homes built prior to 1990 has undergone some renovation, in the rural states of Uri and Nidwalden, in central Switzerland, only 38 per cent of the equivalent housing stock has been renovated. For multi-apartment housing constructed prior to 1990, the renovation rate varies between 67 per cent (the rural canton of Jura) and 24 per cent in Geneva (Konferenz Kantonaler Energiedirektoren, 2014).

'Switzerland is a country of tenants' as the saying goes. It has the lowest rate of home ownership and consequently the highest percentage of tenants in Europe. This is due to high land prices and construction costs, the absence of publicly supported saving schemes for future home owners, and possibly a less pronounced 'my home is my castle' ideology, compared to other European countries. According to Eurostat statistics, on average 58.1 per cent of Europeans (EU28) live in detached or semi-detached homes, compared to only 36.8 per cent of the Swiss population, with almost 60 per cent of the latter living in flats. Although around 53 per cent of Germans and 65 per cent of Spanish residents also live in apartments, what distinguishes these two groups from Swiss residents is ownership. Home ownership is relatively low compared with the other countries under examination here, at 38 per cent for single-family homes and condominiums, although this signals an increase from 31 per cent in 1990. In urban cantons and particularly in cities, the percentage is even lower. While, for example, ownership in 2013 had increased to 27.4 per cent in the canton of Zurich, in the city of Zurich it amounted to only 8.1 per cent. Ownership also varies considerably with age. While only 26 per cent of young and middle-aged households (25–64) owned their home in 2013, the ownership rate among older households (65+) was 47.6 per cent (Swiss Statistics, 2013). It is worth noting that 63 per cent of Swiss single-family homes are located in urban areas, compared to only 22 per cent in Germany (Itard and Meijer, 2008: 32).

Kemeny (1995) classifies Switzerland within the 'unitary rental' model. Gruis and Van Wezemael (2006), comparing the Dutch and the Swiss unitary rental markets, note important differences between the two, however. While the main providers in the Swiss rental market are commercial private landlords, not-for-profit social landlords own the largest share of Dutch rental dwellings. Owning around 60 per cent of Swiss rental housing, the private landlords have typically inherited or acquired larger or smaller housing complexes for investment purposes.

Another 20 per cent of rental housing is owned by large institutional investors, such as pension funds, life insurance companies and banks. The remaining rental stock is in the possession of municipalities, foundations and non-profit housing cooperatives. The share of non-profit housing cooperatives amounts to only 5.1 per cent in Switzerland. The percentage is much higher in cities, however, particularly Zurich and Biel, where this type of housing stock makes up almost one-fifth of all rental housing (Swiss Statistics, 2004). This is highly significant for sustainable housing, as will be discussed later.

In contrast to other countries, particularly in Southern Europe, the last housing crisis in Switzerland dates back to the early 1990s, when a speculative housing investment bubble burst (Pfiffner, 2011). Demand for housing was high before and after the 2008 sub-prime crisis, which barely affected the Swiss housing market, although it did have an impact on some of the large banks. Ongoing high demand is reflected in persistently low vacancy rates, especially in urban regions. A functioning housing market presupposes a vacancy rate of not less than 1 per cent. In the urban cantons of Zurich, Basel, Geneva and Zug, it was below 0.5 per cent in 2015 (Swiss Federal Housing Office, 2015b). In major cities it was lower still, with only 0.11 per cent in Zurich and 0.4 per cent in Geneva and Berne (Swiss Estates, 2016). Only in 2014 was there a significant increase in new available housing. Between 2002 and 2013, the average number of new apartments built per year increased from 29,000 to 47,000 units (Swiss Federal Housing Office, 2015a). The new apartment mix included smaller apartments, reflecting the demand of single or couple households which currently make up about two-thirds of households in the larger cities.

Housing prices in Switzerland vary greatly, not only between cities and surrounding suburban communities, but also among cities. A comparison of monthly rental costs for apartments advertised between January 2013 and March 2015 showed that an equivalent apartment cost twice as much in Zurich as in the more remote city of La Chaux-de-Fonds. Housing prices are highest in Zurich and Geneva, followed by Winterthur (near Zurich) and Lausanne (near Geneva). In 2015, rental costs in Zurich and Geneva were 30 and 33 Swiss francs per m^2/month, respectively (Comparis.ch, 2015). These figures do not include additional costs for heating, water, electricity and maintenance. On average, the latter make up 12 per cent of the rent and they are added to the monthly payments. Heating in older housing can account for about half of these additional costs (Credit Suisse Research, 2006).

The high price of new housing is also due to increased floor-space consumption per capita, a result of changing household compositions, affluence and lifestyle changes. In 2013, the per capita floor-space consumption was $41m^2$ in Zurich, compared to $45m^2$ nationwide. This represents an increase of 40 per cent since 1970 (Stadt Zürich, 2013). Land and residential property prices have also increased significantly due to high demand. In the canton of Zurich, for example, land costs increased by 20 per cent from 2005 to 2011 (Kälin, 2011). This was fuelled by very low mortgage rates, which dropped from around 5.5 per cent in 2000 to between 2 and 2.5 per cent in 2010 (Finanzmonitor, 2011), and remained low until 2015. At the same time, the index for rental costs has risen only mode-rately, with an annual increase of 1 per cent. This is due to practically zero inflation in Switzerland since 2008, as well as the low mortgage interest rate, which is used as a reference for determining rental prices in existing buildings (Swiss Federal Housing Office, 2015b).

Sustainable communities and housing in Switzerland

To be sustainable, community and housing development has to encompass the three pillars of environmental, social and economic sustainability. A key focus of Swiss national and cantonal sustainable development strategies and policies is on environmental protection and energy-related aspects.

Environmental protection, energy production and consumption policies

According to the Environmental Protection Index, in 2014 Switzerland ranked highest among European countries with a score of 87.67, compared to the EU28 average of 72.36 (see Table 1.1). Similarly, a recent Sustainable Government Indicators (SGI) survey ranked Switzerland second among OECD and EU countries in terms of environmental policies (SGI, 2015). These positive rankings are partly a result of national and local policies and intervention measures established since the 1990s. Switzerland's economic development – in contrast to other European countries – has never relied on much large-scale smoke-stack industry because of decentralised industrial development, the lack of natural resources such as coal, and so on.

The SGI (2015) report on Switzerland highlighted the country's achievements in a number of cross-sectoral strategies implemented in recent years. They address biodiversity, climate-change adaptation, forest and natural water resource management, in addition to controlling water pollution. Public spending on environmental protection measures in Switzerland was about 2.5 per cent of total public expenditure in 2012. Furthermore, large investments have been made to improve the railway infrastructure, particularly with regards to trans-alpine freight options, as well as local public transportation. Switzerland's recycling rate is one of the highest in the world. However, although air quality has improved over the past twenty-five years, threshold values of ozone and other gases are still frequently exceeded in summer. According to the report, there has been limited success in terms of nature conservation and landscape protection as the number of animal and plant species that have become extinct or are at risk of extinction continues to rise. Additionally, little progress has been made with respect to controlling noise pollution, with 25–30 per cent of the population still exposed to high levels of noise from road and rail traffic. The report concludes that global environmental policy is high among Switzerland's foreign-policy priorities and that the country has played a significant role in designing and advancing global environmental-protection regimes (SGI, 2015).

Following the Rio Declaration of 1992, the Swiss government decided to make sustainable development an overarching national policy goal. In this context, the Swiss government has regularly published a strategy report since 1997. As a tool for monitoring progress toward sustainability, an encompassing set of principles or indicators – the so-called 'Monet Indicators' – was developed in 2003.

> The principles indicate the direction to be taken in order to create and maintain sustainable development. They form the frame of reference that is used to assess the sustainability of observed developments. The principles are assigned to the three goal dimensions of social solidarity, economic efficiency and ecological responsibility, and they are classified into 20 areas. They provide information about meeting needs and preserving stocks of capital, and also about the efficiency and fairness with which needs are met and resources are used.
>
> (Swiss Statistics, 2016b: 1)

A simplified set of thirty indicators – the so-called 'Cercle Indicateurs' (Indicator Circle) – is now used to measure cantons' and municipalities' sustainability performance both comparatively and over time (Swiss Federal Office for Spatial Planning, 2015).

The most recent sustainable development strategy report encompasses the time period 2012–2015 (Swiss Federal Council, 2012). The introduction emphasises the need to learn from existing successful projects and policies. It also points to the importance of newly developing and existing neighbourhoods in the context of urban design and planning. Neighbourhoods or local communities are considered focal points of public discussion and are seen as the places where sustainability principles in terms of energy-efficient buildings, sustainable mobility strategies and social integration and quality of life for all population groups can best be implemented. Table 11.1 lists the ten key areas for which action plans are to be developed at different levels in terms of national, cantonal and local policies (Swiss Federal Council, 2012).

Table 11.1 Key areas for action: 2012–2015 sustainable development strategy

1. Climate change
2. Energy
3. Spatial development and traffic
4. Economy, production and consumption
5. Use and protection of natural resources
6. Social cohesion, demography and migration
7. Public health, sport and illness prevention
8. Global development and environmental challenges
9. Finance management and provision of incentives
10. Education, research and innovation

Source: Swiss Federal Council, 2012.

Climate change and energy are at the top of the public agenda. The objectives for these areas are: climate protection to prevent irreversible global and local consequences; and reduction of energy consumption and promoting the use of renewable energy sources (Swiss Federal Council, 2012). Current strategies related to climate protection build on a federal (CO_2) law, first passed in 1999 in the wake of Switzerland signing the Kyoto Protocol. In 2009, the Swiss government revised the CO_2 law, stipulating a reduction of CO_2 emissions between the years 2013 and 2020 by at least 20 per cent compared to 1990 levels. Earlier measures focused on the reduction of CO_2 emissions by private cars, as well as integral risk management. They were to be complemented as of 2012 by a mix of climate-change mitigation measures aimed at reducing CO_2 emissions through taxes, trade of emission certificates, subsidies and CO_2 emission restriction laws.

According to the 2014 global ranking by the World Energy Council, Switzerland and Sweden were the only countries with a AAA rating in the Energy Trilemma Index. This sustainability index encompasses the three dimensions of energy security, energy equity and environmental sustainability (World Energy Council, 2014). The Swiss sustainability strategy for 2012–2015 stipulates that energy production and consumption must focus on sustainable management of energy sources aimed at meeting the needs of the economy and society through more efficient energy use and the support of renewable energy sources. The ongoing 'Energy Switzerland' programme involves a variety of measures implemented through cooperation among the federal government, the cantons, the municipalities, and environmental and consumer organisations (Swiss Federal Council, 2012). While working partnerships between the different levels of government work quite well, opposition comes from various interest groups, such as the auto lobby and the nuclear industry.

In 2010, primary energy used in Switzerland encompassed fossil fuels (44.8 per cent), nuclear energy (23.1 per cent), hydropower (11.3 per cent) and natural gas (10.6 per cent). Of the 10.2 per cent from other energy sources, the combined total share of solar, wind, biogas and geothermal sources amounted to only 1.4 per cent (Swiss Federal Council, 2012). The electricity generation mix includes hydropower (56 per cent) and nuclear power (38 per cent), with the remaining 6 per cent drawing on waste incineration, as well as renewable energy sources, such as solar, wind and biomass (Swiss Federal Office of Energy, 2015).

The Swiss Energy Strategy 2050

An important shift in the national energy policy with a far-reaching impact occurred in 2011 in the wake of the Fukushima nuclear disaster. As part of the Energy Strategy 2050, the Swiss parliament and Federal Council decided on an exit policy related to nuclear energy, with the five existing nuclear power plants to be phased out gradually (*Handelszeitung*, 2011). Mühleberg, the first of the old plants, will be idle as of 2020. This nuclear energy exit strategy requires major efforts to reduce electricity consumption and promote alternative energy sources.

This affects the building/housing sector, which consumed almost 50 per cent of the total demand of primary energy required in 2013. Thirty per cent was used for heating, cooling and warm-water generation; 14 per cent for electricity; and 6 per cent for building construction and maintenance. By far the biggest share of energy primarily drawn from fossil fuels (oil heating) is consumed by the housing stock. In 2012, 47 per cent of all single-family homes had an oil heating system, 12 per cent used electricity, and 26 per cent used renewable energy sources. In multi-family homes, 57 per cent used oil heating systems, 7 per cent used electricity, and 16 per cent utilised renewable energy sources. In single-family homes built after 1990, the share of renewable energy use has increased to 44 per cent (Konferenz Kantonaler Energiedirektoren, 2014: 7).

As mentioned earlier, Switzerland has a relatively large number of old, not yet renovated residential buildings, which poses a challenge to improving environmental sustainability of the total housing stock. Energy conservation measures related to housing have thus become part of a national building programme. It is focused on the reduction of CO_2 emissions through a mixed approach of charging CO_2 emission taxes for the use of fossil fuels and redistributing them to subsidise direct measures to reduce CO_2 emissions. One-third of the revenue from the total CO_2 taxes (approximately 450 million Swiss francs in 2013) is returned to the cantons as general contributions. More than half of these contributions went into subsidising renewable energy facilities such as wood heating systems, solar panels and other environmentally sound heating sources (Nicol *et al.*, 2012).

Energy efficiency standards in housing referred to by the cantons are established by the Swiss Association of Architects and Engineers (SIA, 2011), the Conference of Cantonal Energy Directors (EnDK), and 'Minergie'. The latter is a certification process focused on energy efficiency. It includes highly insulated building skins, biomass pellet/wood heating, ventilation and heat recovery, as well as various additional technical installations, such as heat pumps, solar cells and photovoltaic systems. Of approximately 33,000 buildings that have attained the Minergie label since it was introduced in 1994, 85 per cent relate to new housing construction and 7 per cent to housing renovation. Buyers are willing to accept 7 per cent higher building costs for a unit with a Minergie label (Konferenz Kantonaler Energiedirektoren, 2014). Other standards about appropriate and efficient use of energy for room heating and hot-water systems are based on the European standard EN ISO 13790 (Nicol *et al.*, 2012). The total of €640 million paid to property owners between 2010 and 2014 led to an estimated CO_2 reduction of 58,000GWh over the lifespan of the measures taken. Most of them relate to roof and facade insulation measures (Konferenz Kantonaler Energiedirektoren, 2014).

Issues of social sustainability relate, on the one hand, to income, unemployment, the welfare regime, taxation and the immigration policies briefly outlined above and will not be discussed further here. Developing sustainable urban communities and housing, on the other hand, is mostly addressed at the local municipal or community level and will be described below.

Barriers, drivers and challenges for sustainable communities in Switzerland

Barriers and challenges to sustainable urban communities are to be addressed at different political levels: national spatial planning and land use policies based on a high degree of cantonal and local planning autonomy related to zoning laws; and high-priced urban housing markets along with renovation requirements of some of the old housing stock, both of which impact on access to housing for lower-income groups. One the one hand, the Swiss political system, with its decentralised democratic decision-making structures, is a barrier to sustainable development. The same participatory options, however, can be a driver for positive change in terms of the challenges outlined above, as the examples below will illustrate.

National spatial planning and land use policies

Sustainable communities and housing in urban settings require, by definition, a high level of density. High building density can enhance urban qualities, if it is accompanied by a high degree of social and functional density. The attractiveness of many European historical city centres is based on these combined characteristics. Sustainable communities will have to define goals again along these qualities, utilising today's means and technologies for sustainable development. More than 75 per cent of the Swiss population lives in the central plain. Due to the mountain ranges, habitable space is only around 40 per cent of the total area of 41,285 square kilometres. Urbanisation, defined as the percentage of urban dwellers in the total population, is quite high at 73.85 per cent in 2013 (Statista, 2016). According to Eurostat definitions, however, only 26.2 per cent of Swiss residents lived in densely populated urban areas in 2012 (Table 1.1). The far-reaching autonomy of the cantons and particularly the local communities in determining land use has created considerable sprawl. Suburban communities, small cities and villages – less so larger cities – have used local zoning laws to set aside much land for single-family home construction. The goal was to expand the local tax base by attracting middle- to high-income households.

In recent years, public awareness has grown that this type of land use strategy is unsustainable in the long run, as habitable space is scarce. Voting decisions by the Swiss public indicate a change in perspective. For example, a revision of the Swiss spatial planning law – proposed by the Swiss Federal Council – was accepted by voters in 2013 with a 62.9 per cent majority (SRF, 2013). The law provides a binding framework requiring communities to focus further development within existing boundaries and settlement areas, thus promoting higher building density (Swiss Federal Office for Spatial Planning, 2014). Additional housing construction should be focused in urban and suburban areas, transforming 'brownfield' sites and creating attractive open spaces (Mombelli, 2012).

Urban housing markets, sustainability and access to housing

According to a scenario presented by Swiss Statistics (2010), metropolitan areas such as Zurich and Geneva are expected to show a continuing above-average growth rate with concomitant pressures on their urban housing markets. In order to prevent gentrification processes and segregation, these urban housing markets face the challenge of not only providing expensive housing for upper-middle-class and high-income groups. Switzerland does not have a national or cantonal policy for the provision of affordable, so-called 'social' housing, as it exists in many other European countries. Finding appropriate housing is thus left to the people themselves and depends on local programmes and options in cities where the housing market is tight.

Disadvantaged groups in the housing market typically include lower-income immigrant groups, students, some elderly people with small pensions, physically handicapped and other persons with psychological and physical health or social problems as well as families and single-parent households with below-poverty-line incomes. In housing markets with extremely low vacancy rates, such as Zurich and Geneva, access to housing has also become a challenge for middle-class households with moderate incomes, particularly young families. According to Eurostat, the housing cost overburden rate (households with more than 40 per cent of disposable household income devoted to housing) seems to be low in Switzerland. It affects only 16.6 per cent of tenants renting at market price, compared to the EU28 average of 26.2 per cent (see Table 1.2). The felt burden also depends on the level of household income, however. In 2000, Swiss households with an income below €3,680 per month paid 33 per cent of their income in rent, on average. The Swiss Federal Housing Office suggests that housing cost which exceeds 25 per cent of income negatively affects a household's ability to meet other basic needs. In 2009, this was the case for around 25 per cent of all Swiss households (Welter, 2012: 23).

Many of these households live in old urban housing stock that might be cheap, particularly if they have lived there for a long time. Tenant protection legislation stipulates that rents cannot be arbitrarily increased, unless major renovations (bathrooms, kitchens, window replacement, facade insulation and so on) have taken place. Some commercial private owners, focused on short-term revenue, show little interest in investing in improving their multi-family housing units. Rents then remain cheap and profits are high, especially as mortgage interest rates have been very low in recent years. This creates a contradiction between affordability, housing quality and environmental sustainability, as these buildings are very inefficient in terms of CO_2 emissions and energy consumption. If necessary renovations are implemented, rents are raised, making renovated apartments unaffordable for some households. Tenants frequently have to move out if a building is to be substantially renovated, and finding an alternative apartment at a similar price often proves impossible. Very tight housing markets entail the risk that some owners take advantage of tenants who have no choice. An example is the widely publicised case of a 'slumlord' in Zurich who rented dilapidated apartments in his buildings at exorbitant rents to immigrants and other tenants with no other options. Broken toilets and cockroaches were only some of the problems (Fassbind, 2015).

Local initiatives for building sustainable urban communities

The challenge of building sustainable communities in urban centres in Switzerland thus has to address the tight housing market due to economic growth, immigration and the increasing attractiveness of urban living. In the absence of a national low-cost housing policy, every growing city has to design its own strategies and implement local policies and programmes in order to counteract such developments.

Within the nationally defined framework of sustainability indicators, the drivers toward building sustainable communities and neighbourhoods are various local policies and programmes, developed largely in bigger cities. They reflect the far-reaching local planning and decision-making autonomy, characterising the Swiss federalist system. The bottom-up political instruments of 'initiative' and 'referendum' can be used in two ways: in the case of a majority vote on an initiative, it can promote and force new policies; in the case of a successful referendum, it can reverse government decisions. This was evidenced in the majority vote on a more sustainable land use policy at the national level mentioned above. It is illustrated below by important examples in the city of Zurich, which is the financial and economic centre of Switzerland as well as the largest city and metropolitan area. The city proper comprised 400,000 inhabitants in 2015, while the metropolitan area had a population of around 1.2 million people. More than an eighth of all Swiss residents therefore live in the region, and every ninth job is located there (Statistik Stadt Zürich, 2012: 483).

Supported by their inhabitants, Zurich and the city of Basel have adopted specific goals and policies toward creating more sustainable communities. These visions are reflected in a long-term strategy toward achieving a '2,000-watt society' by reducing the current 6,000 watt per capita energy consumption in Switzerland by two-thirds (Palmer, 2009). Both cities have anchored this goal in their governing policies as a mandate by their residents. Zurich voters decided in 2008 to include the goal of reaching a 2000-watt society by 2050 in the city's constitution. This ambition involves a variety of measures in the areas of energy efficiency and renewable energy; requirements for sustainable buildings; mobility for the future; and increasing public awareness (Stadt Zürich, n.d.). At the same time, the vision seeks to enhance the manifold qualities of urban life for the inhabitants. Basel similarly launched the concept of the 'Pilot Region Basel' in 2001. The programme involves partnerships among industry, universities, research institutes and the city authorities, with demonstration projects related to energy-efficient buildings, electricity generation from renewable energy sources, and vehicles using natural gas, hydrogen and biogas. The aims are to put research into practice, seeking continuous improvements, and to communicate progress to all interested parties.

In addition to the long-term strategies outlined by the 2000-watt society projects, major cities develop their own guidelines to support sustainability in a variety of specific areas. The city of Zurich, for example, has published 'guidelines on 4 success factors for socially sustainable housing renovation and new housing construction' (Stadt Zürich, 2015). This non-binding policy initiative addresses the various city departments that are involved in housing renovation and construction, as well as commercial private and larger institutional owners and developers of housing, including non-profit housing cooperatives. The four defined success factors are:

• affordable rental costs due to cost-reduction measures and occupancy guidelines reducing per capita space consumption;
• long-term renewal or renovation strategies with early communication to tenants aimed at encouraging lively, socially mixed neighbourhoods;
• support of a sense of identity and community through the provision of communal indoor and outdoor spaces and infrastructure services, such as childcare; and
• a variety of apartment mix, long-term flexibility and adaptability of use aimed at a high quality of life for tenants of different ages, origins, household types and incomes.

Furthermore, in 2012, the city of Zurich introduced an award competition for completed sustainable renovation projects. An expert jury selected a total of fourteen individual multi-family buildings as well as larger housing complexes for exemplary environmentally and economically sound and socially sustainable housing renovation (Stadt Zürich, 2015). These strategies are designed to encourage housing owners, investors and developers to approach housing construction and renovation in a sustainable manner.

In Zurich, efforts toward developing sustainable neighbourhoods are supported by the high percentage of non-profit housing owned by the municipality and housing cooperatives. They are based on a political vision that non-profit housing should be not only affordable but environmentally sustainable and of high architectural quality. Zurich's non-profit housing accounts for 25 per cent of the city's total housing stock, compared to a very modest 5.1 per cent non-profit housing share nationwide (Swiss Statistics, 2004). A 76 per cent majority of Zurich voters accepted a bottom-up political initiative in 2011 (based on the submission of at least 3,000 valid signatures) to increase this share from one-quarter to one-third by 2050. This goal is to be reached by constructing additional city-owned rental housing and supporting new housing construction by non-profit cooperatives. One way to encourage sustainable non-profit housing construction is for the city to lease city-owned land to interested parties for a period of seventy–ninety years against an annual interest payment. This is an attractive option for housing cooperatives, as it requires less investment capital. City government, in turn, representing the broader public interest, keeps control over the land in order to decide on its use based on future need. For new housing to be built on public land, projects need to be based on a planning/design competition. City experts are members of the design competition jury. The city can thus influence projects in terms of ecological and energy-efficiency concerns as well as social aspects, such as spatial programme, apartment mix, functional use and so on. Pursuing such goals is also in the interests of housing cooperatives, as the examples in the section below illustrate.

Innovative housing projects in Zurich: The role of non-profit housing cooperatives

This section highlights some innovative examples of future-oriented sustainable housing and community development. The city of Zurich and some non-profit housing cooperatives have become innovation leaders in renovating existing stock and building new housing in a highly sustainable way. Projects not only focus on high energy efficiency and reduced environmental impact (i.e. including mobility concepts to reduce or proscribe the use of private cars by tenants) but also include broader neighbourhood development and access to housing for disadvantaged population groups. The examples discussed below include a housing renewal project and two housing complexes that were completed between 2013 and 2015.

Housing renewal: Sihlfeld – a flagship of the 1920s made fit for the future

The Sihlfeld housing complex is owned by Allgemeine Baugenossenschaft Zurich (ABZ) – the largest Swiss housing cooperative, with a total of more than 4,500 apartments – which was founded in 1916. Much of ABZ's housing stock is therefore quite old, although it also owns a few new innovative housing complexes that have been built since the late 1990s. ABZ's management developed a long-term strategy for renovating or replacing its older buildings. The renewal of Sihlfeld, built in 1925, illustrates that replacement is not always an appropriate way to deal with old housing stock. Demolition would not have been allowed as the housing complex is under historic monument protection, but it would also have made no sense. Before the conversion, there were 147 apartments, two-thirds of which were small units with two bedrooms or less. Tenant mix had become a problem, with mostly older people living there. At the same time, apartment access was not barrier-free as there were no elevators in the five-storey buildings. Apartments were increasingly unattractive to families, since rooms were small and floor plans outdated.

The renewal project added six generously sized family apartments to the previous attic floor. Other small apartments were changed and enlarged, with terraces opening toward the inner courtyard. As a result, the number of children living in Sihlfeld almost doubled, thus achieving the goal of a new tenant mix. Elevators were added to some of the staircases, allowing older people with mobility restrictions to stay in barrier-free apartments. Building renewal also

Photo 11.1 Sihlfeld: inner courtyard with new family apartments and generous balconies.
© Angelika Marxer; permission granted.

reduced energy consumption and CO2 emissions. The former oil-fired heating was replaced by geothermal energy, and solar cells on the roof were installed to heat water (Hugentobler, 2014).

New build: Kalkbreite – a new piece of city

In 2006, a handful of residents in Wiedikon – an older, centrally located Zurich neighbourhood – together with housing experts and activists started to develop their vision of Kalkbreite. A new, non-profit housing cooperative of the same name was established. The project was to be built on 6,350 square metres of city-owned land, offered to the housing cooperative on a ninety-two-year lease against an annual interest payment.

Kalkbreite would establish a socially and ecologically sustainable urban living space; a housing and commercial complex featuring contemporary architecture. The principal objective was to provide affordable rents for a balanced mix of residents, varying in household composition, age, gender, income and wealth, as well as education level and nationality. Furthermore, households disadvantaged in the free housing market should be included, but not represent a majority. Collaborations with a variety of social institutions were therefore established in

Photo 11.2 Sihlfeld: renovated, monument-protected facade with newly added penthouse apartments.

© Hannes Henz; permission granted.

the early planning phase. Among the partners was a non-profit foundation, which assists low-income, mostly migrant families in finding decent and affordable housing. Residents from seventeen nations found new homes in the housing complex. Fifteen per cent of the apartments are subsidised by contributions of the canton and reserved for households below a specified income level. Within the cooperative, a solidarity fund supported four other households with limited incomes in 2015.

Traditional living spaces – ranging from small studios to five–six-room apartments – were to be complemented by more experimental new apartment types. The latter included so-called cluster apartments, where individual rooms with a private shower/toilet and a small kitchen are connected to a shared large cooking/dining/living space. Furthermore, several apartments of different sizes were joined and connected to a large communal kitchen and living space, providing living space for several families with up to seventeen rooms. The spatial programme that determined the apartment mix and additional functions the complex was to serve was developed with the participation of interested individuals, who became members of the newly founded housing cooperative. Participation and dialogue started in the early planning phase and continued during the construction until occupancy began in 2014. As a result, regular meetings and participation continue to be part of everyday life in the facility's operating phase.

Kalkbreite encompasses a total of 97 apartments, 33 commercial units, and several rooms of various sizes for common use by the tenants. The ground floor provides space for a variety of shops, including an organic food store, offices, restaurants and a cinema. This mixed use adds to the basic infrastructure and enhances the liveliness of the neighbourhood. Zurich's main train station can be reached in six minutes by bus or tram, with stops located right outside the complex. The building also houses a tram depot on the ground floor, which allows for dramatic spaces with high ceilings at street level. As the housing complex is surrounded by three very busy streets and a train track, apartments are oriented toward an interior courtyard with stairs leading to a roof terrace that is jointly used by the tenants.

Kalkbreite is also exemplary in terms of ecological and energy-efficient building design and low individual energy use. The average floor-space consumption per person is just 32 square metres, due to occupancy guidelines which specify that the number of rooms in an apartment minus one equals the number of minimal inhabitants. Also, none of the inhabitants owns a car (see discussion of the 2000-watt society, above).

'More than housing' – building a new neighbourhood

Another housing project on a much larger scale was initiated in 2007, on the hundredth anniversary of non-profit housing in Zurich. It encompasses more than 300 apartments in 13 buildings that provide attractive living space for around 1,400 people. More than fifty Zurich non-profit housing cooperatives engaged in

Photo 11.3 Kalkbreite: streetside with shops and cinema.

© Volker Schoop; permission granted.

Photo 11.4 Kalkbreite: inner courtyard and roof terrace.

© Volker Schoop; permission granted.

this project, together with foundations, social service institutions, the city of Zurich (which provided the land on a lease basis) and the Swiss Federal Housing Office. 'More than housing' was completed and inaugurated in the summer of 2015, with the first occupants arriving in the fall of 2014. Goals are similar to those of Kalkbreite. For example, various apartment options were created for different household types and tenants, including immigrant families as well as people with disabilities. The mix of infrastructure services provided on the ground floor includes: childcare facilities, space for communal use, shops, restaurants and so on. Per capita floor-space consumption has been limited, car ownership restricted and inhabitants have been encouraged to participate in decision-making about how to use the outdoor space, as well as many other tenant-driven initiatives (Hugentobler *et al.*, 2016).

Conclusions

In Switzerland, the characteristics of rental housing, largely owned by many small private landlords, as well as the tight urban housing market driving up costs and limiting access, negatively affect the sustainability of communities and urban housing. By contrast, government initiatives at the national level, such as the sustainable development strategy 2012–2015 and particularly the Energy Strategy 2050, are promising steps toward a more sustainable future for urban communities. They involve a variety of specific programmes, standards and tools toward sustainable housing construction and renovation as well as supporting more sustainable resource consumption. Many of these programmes relate to housing as a major sector of energy consumption and CO_2 emissions and also to efforts to support more sustainable communities, enhancing urban as well as individual quality of life.

In terms of local implementation and overall long-term progress, overarching initiatives such as the 2000-watt society programmes in the cities of Zurich and Basel are even more promising. They were initiated bottom-up and/or by local executive governments with social democratic and green party background majorities. A progressive urban voting population supports these policies. The particular democratic decision-making structures, where voters can influence decision-making at all levels of government, thus allow for innovative local as well as national initiatives. Examples are the majority vote on increasing the share of non-profit housing in Zurich locally, and the acceptance of a revised spatial planning law at the national level.

Comparative European statistics on per capita GDP, unemployment, risk of poverty or social exclusion (Table 1.1) indicate that Switzerland is a wealthy country. With overall high-quality housing, excellent public transportation and many thriving urban areas, one might ask what problems the Swiss face. However, almost half a million people in Switzerland live below the poverty line (Swiss Statistics, 2012) and know some of these problems first-hand. National statistics often obscure the hardship of minorities. As outlined in the section on barriers, drivers and challenges above, access to affordable and decent housing in major Swiss cities is a major problem for students, older people on low incomes, single-parent households and refugees and immigrants with low educational and professional qualifications. These barriers to sustainable and inclusive local community development will remain an issue in ongoing political discussion, particularly in the major cities. The Swiss Federal Housing Office (2015a) describes the challenges more optimistically: 'Energy-related housing renewal along with further price increases, the necessary adaptation to the needs of the aging population, and evolving social needs create significant potential for innovation and new models in projects and processes.'

In a globalising world, Switzerland is no island. Economic growth slowed in 2015 in the wake of an unexpected Swiss National Bank (SNB) decision in January of that year. Until then, the SNB had actively intervened in the market in order to sustain a stable exchange rate of 1.2 Swiss francs to the euro. Abandoning this increasingly expensive policy, the euro weakened considerably against the Swiss currency, making the importation of Swiss goods more expensive for Switzerland's very important Euro Zone trading partners. The unemployment rate is expected to increase as exports decline and some local production has already shifted abroad. Furthermore, the acceptance of the populist-driven 'immigration initiative' by Swiss voters in 2012 – by a narrow margin of 50.2 per cent – presents an as yet unresolved problem. It threatens all bilateral agreements between Switzerland and the EU if a solution cannot be negotiated. And while the number of refugees from war-torn countries arriving in Switzerland since mid-2015 has been negligible in comparison to the tens of thousands accepted by Italy, Greece and Germany, it fuels populist sentiment amid arguments that this major challenge for Europe will hardly spare Switzerland.

Furthermore, the considerable gains of the Swiss Populist Party (SVP) in the parliamentary elections of October 2015 – largely at the expense of green and the green–liberal parties – are likely to threaten not only liberal immigration policies but also the implementation of the Energy Strategy 2050, which advocates the phasing out of nuclear energy in Switzerland. This conservative shift may also have a negative impact on the future of sustainable local communities. As Peter von Matt, a well-known Swiss author and professor of German literature, commented on television when discussing the election results: 'Switzerland still has a strong bomb-shelter mentality in the face of external threats.'

In summary, economic and political developments may well threaten current national efforts toward building sustainable communities. It can only be hoped that the downside of the Swiss democratic system – reflected in voters' acceptance of several populist initiatives over recent years – will be counteracted by the same democratic decision-making structures. Future-oriented perspectives are needed that support progressive city government policies toward sustainable communities and inclusive urban housing policies.

References

All websites accessed 13 July 2016 unless otherwise stated.

Armingeon, K., Bertozzi, F. and Bonoli, G. (2004) 'Swiss worlds of welfare', *West European Politics*, 27: 20–40.

Arts, W. and Gelissen, J. (2010) 'Models of the welfare state', in Castles, F., Leibfried, S., Lewis, J., Obinger, H. and Pierson, C. (eds) *Oxford Handbook of the Welfare State*. Oxford: Oxford University Press, pp. 569–585.

Bogedan, P., Gindulis, E., Leibfried, S., Moser, J. and Starke, P. (2010) *Transformations of the Welfare State: Small States, Big Lessons*. Oxford: Oxford University Press.

Building Performance Institute Europe (2011) *Europe's Buildings under the Microscope: A Country-by-Country Review of the Energy Performance of Buildings*. Available at www.bpie.eu/uploads/lib/document/attachment/21/LR_EU_B_under_microscope_study.pdf.

Comparis.ch (2015) *Städtevergleich 2015: So teuer ist Wohnen in der Schweiz!* Available at www.comparis.ch/immobilien/news/2015/04/mietpreise-staedte-schweiz-vergleich.aspx.

Credit Suisse Research (2006) *Swiss Issues Immobilien: Die Mieten in der Schweiz*. Available at www.mischol.ch/kundenser/Die%20Mieten%20in%20der%20Schweiz_CS-Studie2006.pdf.

Credit Suisse Research (2015) *Global Wealth Report 2015*. Available at https://publications.credit-suisse.com/tasks/render/file/?fileID=F2425415-DCA7-80B8-EAD989AF9341D47E.

Ecopop (2014) *Bevölkerungsdichte*. Available at www.ecopop.ch/de/bevoelkerung/bevoelkerung-schweiz/453-bevoelkerungsdichte.

Fassbind, T. (2015) *Kakerlaken und kaputte WCs – Wohn-Albtraum mitten in Zürich*. Available at www.tagesanzeiger.ch/zuerich/stadt/razzia-wegen-mietzinswucher/story/18397798.

Finanzmonitor (2011) *Historische Zinsentwicklung Hypotheken*. Available at www.finanzmonitor.com/immobilien-hypothek/zinsentwicklung-hypotheken/.

Gruis, V. and Van Wezemael, J. (2006) 'The Dutch and the Swiss: Two unitary rental markets compared', in *Book of Abstracts: 13th Annual European Real Estate Society Conference*. Weimar: ERES Conference.

Handelszeitung (2011) 'Geschichte der Schweizer Atomkraft', 25 May. Available at www.handelszeitung.ch/konjunktur/schweiz/geschichte-der-schweizer-atomkraft.

Hugentobler, M. (2014) 'Densification: Housing Replacement and Reconstruction', *International Forum of Towns Graz*, 3: 22–26.

Hugentobler, M., Hofer, A. and Simmendinger, P. (eds) (2016) *More than Housing: Cooperative Planning – a Case Study in Zurich*. Basel: Birkhaeuser.

Itard, L. and Meijer, F. (2008) *Towards a Sustainable European Housing Stock: Figures, Facts and Future*. Amsterdam: IOS Press BV.

Kälin, A. (2011) 'Bauland wird rarer und teurer', *Neue Zürcher Zeitung*, 28 September. Available at www.nzz.ch/bauland-wird-rarer-und-teurer-1.12695476.

Kemeny, J. (1995) *From Public Housing to the Social Market: Rental Policy Strategies in Comparative Perspective*. London/New York: Routledge.

Konferenz Kantonaler Energiedirektoren (2014) *Energieverbrauch von Gebäuden*. Available at www.endk.ch/media/archive1/aktuelles/20140828_FactSheet.pdf.

Mombelli, A. (2012) *Neues Leben auf alten Industriearealen*. Available at www.swissinfo.ch/ger/raumplanung_neues-leben-auf-alten-industriearealen/33368954.

Nicol, L. A., Hugentobler M. and Van Wezemael, J. (2012) 'Switzerland: non-profit housing sector – a leader in energy initiatives', in Nieboer, N. (ed.) *Energy Efficiency in Social Housing Management*. London: Earthscan, pp. 134–150.

Novatlantis.ch (n.d.) *Region Basel*. Available at www.novatlantis.ch/region-basel/.

OECD Better Life Index (2015) *Switzerland*. Available at www.oecdbetterlifeindex.org/countries/switzerland.

Palmer, E. (2009) 'Can a city cut its energy use by 2/3?', *CBS Evening News*, 6 December. Available at www.cbsnews.com/news/can-a-city-cut-its-energy-use-by-2-3/.

Pfiffner, F. (2011) '40 Milliarden einfach "verbrannt"', *Neue Zürcher Zeitung*, 27 November.

Schweizerische Eidgenossenschaft (2013) *Bundesrat verabschiedet Botschaft zur Energiestrategie 2050*. Available at www.news.admin.ch/message/index.html?lang=de&msg-id=50123.

SGI (2015) *Switzerland: Environmental Policies*. Available at www.sgi-network.org/2015/Switzerland/Environmental_Policies.

SRF (2013) 'Klares Ja zum Raumplanungsgesetz', *Schweizer Radio und Fernsehen*, 3 March.

Stadt Zürich (2012) *Auszeichnung nachhaltig sanieren*. Available at www.stadt-zuerich.ch/prd/de/index/ueber_das_departement/medien/medienmitteilungen/2012/april/120417a.html/.

Stadt Zürich (2013) *Wie viel Wohnraum braucht der Mensch?* Available at www.stadt-zuerich.ch/prd/de/index/statistik/publikationen-angebote/publikationen/webartikel/2013-03-28_Wie-viel-Wohnraum-braucht-der-Mensch.html.

Stadt Zürich (2014) *Sustainability Monitoring in the City of Zurich: Summary 2014*. Available at www.stadt-zuerich.ch/prd/de/index/stadtentwicklung/stadt-_und_quartierentwicklung/nachhaltige_entwicklung/nachhaltigkeitsmonitoring.html.

Stadt Zürich (2015) *Leitfaden Erfolgsfaktoren sozial nachhaltiger Sanierungen und Ersatzneubauten*. Available at www.stadt-zuerich.ch/nachhaltigsanieren.

Stadt Zürich (n.d.) *2000-Watt Society*. Available at www.stadt-zuerich.ch/portal/en/index/portraet_der_stadt_zuerich/2000-watt_society.html.

Statista (2015) *Anzahl der Erwerbstätigen in der Schweiz nach Wirtschaftssektoren von 2004 bis 2014*. Available at http://de.statista.com/statistik/daten/studie/216757/umfrage/erwerbstaetige-nach-wirtschaftssektoren-in-der-schweiz/.

Statista (2016) *Urbanisierungsgrad in der Schweiz von 2004 bis 2014*. Available at http://de.statista.com/statistik/daten/studie/216770/umfrage/urbanisierung-in-der-schweiz.

Statistik Stadt Zürich (2012) *Statistisches Jahrbuch der Stadt Zürich*. Available at www.stadt-zuerich.ch/prd/de/index/statistik/publikationen-angebote/publikationen/Jahrbuch/statistisches-jahrbuch-der-stadt-zuerich_2012.html.

Statistik Stadt Zürich (2015) *Gebäude und Wohnungen*. Available at www.stadt-zuerich.ch/prd/de/index/statistik/themen/bauen-wohnen/gebaeude-wohnungen.html.

Swiss Association of Architects and Engineers (SIA) (2011) *SIA Energy Efficiency Path*. Available at http://shop.sia.ch/normenwerk/architekt/sia%202040/e/D/Product, accessed 8 August 2016.

Swiss Estates (2016) *Wohnimmobilienmarkt Schweiz*. Available at www.swiss-estates.ch/portrait/immobilienmarkt-schweiz

Swiss Federal Council (2012) *Strategie Nachhaltige Entwicklung 2012–2015*. Available at www.are.admin.ch/dokumentation/publikationen/00014/00399/index.html?lang=de.

Swiss Federal Housing Office (2015a) *BWO-Forschungsprogramm 2016–2019. Beilage 2*. Available at www.bwo.admin.ch/themen/wohnforschung/00163/index.html?lang=de, accessed 8 August 2016.

Swiss Federal Housing Office (2015b) *Leer stehende Wohnungen.* Available at www.bwo. admin.ch/dokumentation/00101/00104/index.html?lang=de.

Swiss Federal Office of Energy (2015) *Schweizerische Elektrizitätsstatistik 2014.* Available at www.bfe.admin.ch/themen/00526/00541/00542/00630/index.html?dossier_id=00765

Swiss Federal Office for Spatial Planning (2014) *Revision des Raumplanungsgesetzes.* Available at www.are.admin.ch/themen/recht/04651/index.html?lang=de.

Swiss Federal Office for Spatial Planning (2015) *Best Practices in Sustainable Development.* Available at www.are.admin.ch/dokumentation/publikationen/00014/000595/index.html ?lang=it&download=NHzLpZeg7t,lnp6IONTU04212Z6ln1ah2oZn4Z2qZpn02Yug 2Z6gpJCEe314g2ym162epYbg2c_JjKbNoKSnbA--.

Swiss Statistics (2004) *Genossenschaftlich Wohnen.* Available at www.bwo.admin.ch/ themen/00328/00331/index.html?lang=de&download=NHzLpZig7t,lnp6IONTU04212 Z6ln1acy4Zn4Z2qZpnO2Yuq2Z6gpJCDdn12hGym162dpYbUzd,Gpd6emK2Oz9aGod etmqaN19XI2IdvoaCUZ,s-..

Swiss Statistics (2010) *Szenarien zur Bevölkerungsentwicklung der Schweiz 2010–2060.* Available at www.bfs.admin.ch/bfs/portal/de/index/news/publikationen.html? publicationID=3989.

Swiss Statistics (2012) *Armut in der Schweiz: Konzepte, Resultate und Methoden. Ergebnisse auf der Basis von SILC 2008 bis 2010.* Available at www.bfs.admin.ch/bfs/ portal/de/index/themen/20/22/publ.html?publicationID=4918.

Swiss Statistics (2013) *Wohnverhältnisse.* Available at www.bfs.admin.ch/bfs/portal/de/ index/themen/09/03.html.

Swiss Statistics (2014) *Gebäude und Wohnungen – Daten, Indikatoren.* Available at www. bfs.admin.ch/bfs/portal/de/index/themen/09/02/blank/key/gebaeude/art_und_groesse. html.

Swiss Statistics (2015a) *Arbeitslosigkeit, offene Stellen – Indikatoren.* Available at www. bfs.admin.ch/bfs/portal/de/index/themen/03/03/blank/key/registrierte_arbeitslose/ entwicklung.htmlSocio-demographic%20developments.

Swiss Statistics (2015b) *Ausländische Bevölkerung: Staatsangehörigkeit.* Available at www.bfs.admin.ch/bfs/portal/de/index/themen/01/07/blank/key/01/01.print.html.

Swiss Statistics (2015c) *Bevölkerungsstand und -struktur – Detaillierte Daten.* Available at www.bfs.admin.ch/bfs/portal/de/index/themen/01/02/blank/data/01.html.

Swiss Statistics (2016a) *Bruttoinlandprodukt – Daten, Indikatoren.* Available at www.bfs. admin.ch/bfs/portal/de/index/themen/04/02/01/key/bip_gemaess_produktionsansatz. html.

Swiss Statistics (2016b) *Sustainable Development – MONET.* Available at www.bfs.admin. ch/bfs/portal/en/index/themen/21/02/ind9.approach.905.html.

Trading Economics (2015) *Switzerland Unemployment Rate.* Available at www.trading economics.com/switzerland/unemployment-rate.

US Internal Revenue Service (IRS) (n.d.) *Translating Foreign Currency into US Dollars.* Available at www.irs.gov/Individuals/International-Taxpayers/Yearly-Average-Currency-Exchange-Rates.

Welter, C. (2012) *"Affordable Housing" als Anlagesegment für institutionelle Investoren?* Available at www.curem.uzh.ch/static/abschlussarbeiten/2012/Welter_Caroline_MT_ 2012.pdf .

World Energy Council (2014) *Energy Trilemma Index.* Available at www.worldenergy.org/ data/trilemma-index/.

12 Germany

Clemens Deilmann and
Karl-Heinz Effenberger

Introduction

The current housing situation in Germany is a result of visions of housing policies and urban planning, the orientation of which has changed over the decades. However, highly stable settlement patterns are influenced by the regional planning principle '*dezentrale Konzentration*' (decentralised centralisation) and by the constitution of the Federal Republic of Germany. The drive towards sustainable communities began in the early postwar decades with the principle of equal opportunity and the objective of social stability. The constitutional rules governing public finances state an important principle: 'Such law shall ensure a reasonable equalisation of the disparate financial capacities of the Länder [federal states], with due regard for the financial capacities and needs of municipalities [associations of municipalities]' (Article 107.2). Strong federal states have to support weak ones financially. Another principle in the constitution stipulates that no law may be passed if it thwarts the establishment of equal standards of living (Article 72.2). Regarding spatial development, this applies especially to the equal availability of infrastructure and development support for small and medium-sized towns. Settlement patterns in Germany are *inter alia* an expression of these two principles. From a satellite perspective, Germany appears rather evenly 'dotted' with settlements. Approximately half of the German population lives in small towns of less than 20,000 inhabitants.

Given this background to spatial planning, strategies to reduce the negative effects of urbanisation on ecology and health have improved living conditions in communities. Since the early 1980s, a number of policies and practices have been developed to demonstrate and stimulate more environmentally friendly sustainable development. These include demonstration cases for green housing, good practice guides, indicator sets, certification schemes, modelling and monitoring tools, new funding schemes and legal frameworks. Today, tools and instruments for urban planning down to the level of the building concretise the national sustainability strategy. Future challenges for the sustainability of communities will arise less from ecological issues and more from social and economic issues due to demographic change, migration and socio-economic divisions. Major concerns include dramatic population loss in East Germany since 1995 and, in the long run, population decline across all of Germany, vacancies in multi-unit residential buildings in East Germany and future vacancies in single-family home areas in West Germany.

Germany had a population of around 80.4 million people in 2011, and by June 2015 this had risen to 81.4 million. Overall, the population has increased slightly since 1991, although between 2005 and 2010 the number of residents decreased slightly. East Germany (including Berlin) recorded a loss of around 2 million people (nearly 12 per cent) between 1991 and 2010. In West Germany, the population increased by approximately 2.1 million (3.4 per cent) over the same period (DESTATIS, 2014). The number of households increased in both East and West Germany between 1991 and 2010. In East Germany, the number of households grew by 10 per cent – from 7.8 million to 8.6 million – despite the declining population. This increase is due to a reduction in the average household size, from 2.31 to 1.88 people per household. In West Germany, the number of households increased by 4.2 million (15 per cent) to 31.7 million, while the number of people per household declined from 2.26 to 2.07. Around two-thirds of the increase in households can be attributed to reduced household size, while one-third results from the increase in population (Effenberger *et al.*, 2014; DESTATIS, 2012). In 2015, GDP for the total economy at current prices is €3,025 billion with a growth rate of 1.7 per cent. Gross monthly earnings were €3,500 in that year (DESTATIS, 2016). There were 42.9 million people in employment, and 2.9 million (6.6 per cent) were registered as unemployed.

Currently, there are 40.4 million residential and non-residential buildings in Germany (DESTATIS, 2013; see Table 12.1). Around 97 per cent of all dwellings are in residential buildings. More than half of these dwellings are in multi-family houses. In East Germany, the share of dwellings in multi-family houses (65.9 per cent) is particularly high. In West Germany, this share amounts to 50.6 per cent, and almost half of all dwellings are single-family or two-family houses. In West Germany, 20.1 per cent of dwellings were built before 1948; in East Germany the figure is 41.9 per cent. The 2011 census recorded more than 1.8 million vacant dwellings in Germany – temporary vacancies resulting from tenancy changeovers are not included in this figure (DESTATIS, 2013). Vacancies have become a threat to the sustainability of communities in some regions and parts of cities, especially in East Germany. In Saxony and Saxony-Anhalt the average vacancy rate is approximately 10 per cent. However, these average figures conceal a dramatic development, as in certain neighbourhoods the vacancy rate can be over 25 per cent. Multi-family housing estates built before 1918 and buildings built between 1979 and 1990 are most affected. After deduction of the vacancies, the inhabited residential area in Germany amounted to about 3,550 million square metres in 2010. Land use for residential purposes is increasing in both parts of Germany (Effenberger *et al.*, 2014).

The housing system in Germany

Even twenty years after the Wall came down, there are still marked differences between East and West Germany with regard to demographic trends and housing

Table 12.1 Development scenarios for housing, East and West Germany, until 2060

East Germany	2010[2]		2011–2020		2021–2030		2031–2040		2041–2050		2051–2060	
	East	West	East	West	East	West	East	West	East	West	East	West
New dwellings (000s)			200	1600	150	1200	130	1100	120	1000	100	850
Reduction (000s)			300	300	400	400	450	800	600	1400	600	1600
Housing stock (000s)[1]	8750	31665	8650	32965	8400	33765	8080	34065	7600	33665	7100	32915
Change in household (000s)			−330	+1075	−225	+465	−295	−75	−540	−890	−420	−1165
Vacancy absolute (000s)[1]	629	1202	859	1427	834	1762	809	2137	869	2627	789	3042
Vacancy rate (per cent)[1]	7.2	3.8	9.9	4.3	9.9	5.2	10.0	6.3	11.4	7.8	11.1	9.2

Source: Effenberger et al., 2014.

Notes: [1] At the end of a decade; [2] Data based on the 2011 census.

stock. In the following sections, it is therefore necessary to mention some of the differences. The contrasting development of stock in East and West Germany started after the end of the Second World War. Severe housing shortages (a housing deficit of nearly 6 million dwellings) made it obvious that the problem could not be solved without significant state intervention (Spiegel, 1978).

In East Germany, only the state took on this task by streamlining mass developments for housing. Property development in the form of single-family houses was marginal. As state-controlled rents were extremely low, it was almost impossible for private owners of pre-war multi-family houses to maintain these buildings from an economic viewpoint. Many handed their buildings over to the state due to economic hardship. The state, however, had no concept of how to maintain or refurbish the older building stock. On the one hand, politics had radically phased out craftsmanship and, on the other, industrialised housing production was unable to solve the various requirements of refurbishment and modernisation of existing housing stock. The limitations of the planned economy were exposed. At the time of reunification, many pre-war buildings were in advanced states of decay or even ruin.

After the Wall came down, at the beginning of the 1990s the German government triggered a residential construction boom in East Germany through an extremely attractive tax relief system, unparalleled in the history of the Federal Republic. It was possible to spread 50 per cent of the investment in new housing construction over one to five years as 'financial loss' in the tax declaration and therefore reduce taxable income considerably. Almost two-thirds of these new constructions were multi-family dwellings. At the same time, many East German households moved to West Germany. By 1998, the vacancy rate had reached more than 12 per cent. Due to the problem of increasing vacant dwellings, the government changed the conditions for residential construction subsidies in 1996 from tax relief to a direct grant of 10 per cent of investment – limited to €140 per square metre. The grants stopped completely in 1999 and additionally a speculation period of ten years was introduced on sales of dwellings. Within the ten-year period tax must be paid on yields. Before this regulation, yields from sales were tax free. In 2002 – due to the continuing high vacancy rates – the federal government took action and introduced the '*Stadtumbau Ost*' (Urban Restructuring in East Germany) programme, which included a demolition subsidy. The demolition of multi-storey buildings was subsidised with €60 per square metre per destroyed residential floor area. The subsidy is still available in some East German federal states.

In West Germany, after the Second World War the state and the private sector formed a partnership regarding housing issues. This partnership was meant to be temporary because the overall goal was a residential construction industry independent of state subsidies. Along with the construction of social housing, the support of private property ownership – owner occupier or landlord – was always a matter of principle for West Germany (Korte, 1972).

Fifty-two per cent of dwellings in Germany are rented. In East Germany, this proportion is 66 per cent. Renting is an important housing choice in Germany.

A strong legal protection of tenants gives security of tenure. Tenants' rights and threats to their status receive prominent media attention and generate political action. Evidence for this is a new law – (BGB) § 556d from June 2015 – which was introduced in reaction to rising prices in the large, attractive cities. Besides the existing rent regulation, which prevents prices from sharp increases in existing tenure contracts, the government introduced the new regulations to dampen rent increases when there is a change in tenant. Under the new law, '*Gesetz zur Dämpfung des Mietanstiegs auf angespannten Wohnungsmärkten*' (Law to Slow the Increase in Rents in Stressed Housing Markets), federal states are free to declare certain areas or cities 'stressed markets' and can limit rent increase when there is a change of tenant to a maximum of 10 per cent above the 'comparable rent' in the city. (Each large city has a rent comparison list, which mirrors the average of existing rents.)

Multi-family houses must be differentiated into types of owners. Of the 3.4 million apartments, 1.4 million (41 per cent) are owned by institutional investors (cooperative building company, municipal housing association or large private owner). Of the 2 million privately owned buildings (around 10 million dwellings), 60 per cent are owned by a single private owner, and 40 per cent are shared among several owners, according to the '*Wohnungseigentümergesetz*' (WEG; Residential Property Act). Thus, around 26 per cent of all dwellings in Germany are owned by private landlords, which is 48 per cent of all dwellings in multi-family buildings and 57 per cent of all rented dwellings (Cischinsky, 2013). About three-quarters of these property owners rent out only one dwelling. The proportion of 'amateur landlords' is very high in Germany. Within the group of institutional owners, there were changes at the beginning of the twenty-first century in the proportions of cooperative, municipal and institutional large private owners. During the privatisation wave at the turn of the century, private (often international) investors bought large portfolios from municipal or cooperative stocks. The proportion of private small owners has remained almost constant.

Supply subsidies

Residential construction consists of: publicly subsidised developments (carried out by non-profit housing associations that also benefit from tax concessions); developments that benefit from tax concessions to construct private properties; and privately financed developments. In West Germany in the 1950s and 1960s, around 500,000 residential units were built each year (in 2010 around 150,000 were built). By 1970, the share of publicly subsidised housing out of all new residential developments had decreased from 60 per cent at the beginning of the 1950s to just 20 per cent. Moreover, the initial focus, in terms of resources, on constructing multi-family houses had shifted towards the construction of single-family houses. Thus, by 1970, 40 per cent of all new residential developments were single- or two-family houses (3 million houses with 5 million dwellings). Of these, a million single-family houses had been subsidised under the framework for social housing scheme (Bauwelt cited in Spiegel, 1978). In 1965, a second

'*Förderweg*' (housing subsidy) was introduced to provide tax concessions for medium-income groups to construct and acquire private property in houses or apartments, and this option absorbed an increasing share of subsidies.

In the 1980s, residential construction programmes increasingly opened up to private investors. The non-profit status and tax concessions for housing associations were abolished. With the construction crisis of the early 1980s, the builder–owner model was invented in order to mobilise private capital. In this model, small private investors joined with a trustee to form builder–owner groups. Prospective tax reliefs attracted private investors. As private capital was now a priority, subsidies for the social housing '*Förderweg 1 and 2*' programmes were completely abolished in 2001 (Egner *et.al.*, 2004). It was not until 2006 that the owner-occupied home allowance – oriented towards single-family houses – was also abolished. Many regional and urban planning experts regarded this as far too late, as this subsidy was viewed as a cause of increasing urban sprawl and undiminished land consumption in Germany. From 2005 onwards, the Kreditanstalt für Wiederaufbau (KfW; Reconstruction Credit Institute) launched new programmes for energy efficiency and regenerative energy production. Further small-differentiated programmes at the federal state level were introduced as the national state withdrew from social housing. The federal states received a total of €500 million per year to support their programmes for modernising residential buildings for certain target groups, and to restructure and construct new residential buildings suitable for the elderly and the disabled. In January 2016, due to the large numbers of migrants and asylum-seekers arriving in Germany, the national government decided to double the subsidies to a total of €1 billion per year. A further increase to €2 billion and tax incentives were being discussed at the time of writing.

Demand subsidies

The Housing Allowance Act, which has been in force since 1964, includes a demand subsidy (*Subjektförderung*). The goal is to enable lower-income groups through rent allowances to rent apartments, offered at an average rental rate in the private housing market. This aims to maintain and create stable and socially mixed neighbourhoods, avoiding social segregation in certain residential buildings/neighbourhoods by enabling people to seek accommodation in the private rental market rather than forcing them to reside in subsidised buildings.

In 2012, around 700,000 households received housing benefit, plus another million with reduced earning capacity or old age minimum income. Sixty per cent of those who receive housing benefits are one-person households, while 38 per cent are households with children. Two-thirds of the 1.7 million households receiving this benefit are out of work, with three-quarters of them pensioners (DESTATIS, 2013). Another 5 million people who receive jobseeker's allowances, social benefits or living-cost support are not included in this figure. The state also covers rent payments for these people, but the benefit is known as rent payment allowance rather than housing benefit. The allowances are staggered according to a number of criteria (household size, income, regional factor, rent level subject to subsidies in the municipality). This allowance lowers the burden of housing costs for the households in need from an average of 36 per cent to 28 per cent of total income (*IVV*, 2015).

There are significant differences in terms of ground-floor plan, interiors, technical fitments, age and location within the housing stock. Large prefabricated housing estates, built from the 1970s in West Germany and in the 1970s and 1980s in East Germany, face challenges regarding their social sustainability. Within these estates, there is often a concentration of welfare recipients and immigrants (GdW WI, 2016). The sustainability of communities is also at stake in small towns, especially those with fewer than 10,000 inhabitants. These municipalities suffer from population loss due to internal migration and demographic change.

Sustainable communities and housing

The introduction of the sustainability concept in public and political discourse initially aimed to integrate ecological issues into the general development concept. It was not until the mid-1990s that sustainability became a frequently used term. The Rio World Conference on Environment and Development (1992) set the guiding principles for environmental policies. Subsequently, many German cities initiated Local Agenda 21 processes and the federal government decided on a twenty-one-indicator monitoring system. Statistische Bundesamt analyses and illustrates performance over time (Federal Statistical Office indicators of the German Strategy for Sustainable Development on the environment and the economy). At the regional and urban planning level, the term 'sustainable community' is rarely used to describe a final envisaged status. Rather, 'sustainable community' is understood as an imperative that demands a continuous process of concretisation. While a general calibration in numbers of what a sustainable community should ultimately look like does not exist, there are nevertheless indicator systems to illustrate development trends towards sustainability. Individual federal states and cities choose which indicators to use. At the neighbourhood level, the term is used in Germany to label outstanding examples of urban development and in some cases reference is made to certification schemes developed by the German Sustainable Building Council. But the basis for the fruitful implementation of sustainability targets was prepared long before that by an intensive discourse on ecological issues between 1970 and 1990.

From ecology to sustainability

The discussion about the limits to growth and the 1973 oil price shock triggered a debate about ecology that continues to the present day. The 1970s also saw demonstrations against nuclear energy and the 'housing struggle' (squatting action) in Frankfurt, which opposed the large-scale demolition of pre-war neighbourhoods. Thinking about alternatives evolved into the 'alternative movement'. This tried to establish an approach that was opposite to 'the unscrupulous domination over nature' and the economic constraints of capitalism through self-organisation and self-sufficiency. It explored all issues of self-sufficiency in terms of food, energy, water cycle, building materials, construction, communication, economy, self-organisation, autonomy and global social justice. The key term 'alternative technology' merges concepts for building and living. The concept of an 'autonomous house' (Vale and Vale, 1975) is propagated. Especially in rural areas, self-sufficient detached houses or settlements were expected to emerge. In this period, Schumacher (1978) defined the new understanding of nature as a cooperative relationship between humans and nature. It is about the principles of preserving diversity, reducing monolithic structures, decentralisation, conserving natural resources, recycling and internalising the external costs of the production of goods. These ideas were viewed as peripheral and outlandish by the majority of West German society in the 1970s. Forty years later, most German citizens accept their validity.

Ecologically sound buildings

The standard volume on 'ecological construction' by Krusche *et al.* (1982), commissioned by the Federal Environmental Agency, dealt with all relevant spheres of activity in the building sector from an architectural point of view. It is a book of guiding principles. In the 1990s, research was conducted across Europe on life-cycle cost-analysis methods (LCA) that can compare ecological designs and products across multiple criteria. This prompted a debate about building assessment tools. BREEAM was the first multi-criteria building assessment tool, and producers and users of office buildings were the first target group for its application. With the assessment, buildings can receive a certificate and thereby gain a competitive advantage over rival buildings. In Germany, a sophisticated yet manageable assessment method has been available only since the launch of the *Deutsche Gesellschaft für Nachhaltiges Bauen* (DGNB) – the foundation of the German Sustainable Building Council – in 2007. The DGNB has developed one of the most ambitious of all internationally known certification systems. The overall performance of buildings in terms of sustainability is assessed on the basis of around forty criteria, including thermal comfort, design for all, sound insulation, but also quality of the design process itself.

> The core system for assessing the sustainability quality of buildings was jointly developed by the DGNB and the Federal Ministry of Transport, Building and Urban Development [Bundesministerium für Verkehr, Bau und Stadtentwicklung; BMVBS] in 2009. While the BMVBS tailored and specified this basis for its own evaluation of federal buildings, the DGNB used it to develop a complete certification system for the most diverse types of building and neighbourhood use.
>
> (DGNB, 2016)

In the meantime, around 400 projects worldwide have been certified according to the DGNB. Application of the system is voluntary, but the federal government has made it obligatory for its buildings. The criteria are listed in Box 12.1.

Box 12.1   Criteria for DGNB certification

Environmental quality

1. Life-cycle Impact Assessment
2. Local Environment Impact
3. Responsible Procurement
4. Life-cycle Impact Assessment – Primary Energy
5. Drinking Water Demand and Waste Water Volume
6. Land Use

Economic quality

1. Life-cycle Cost
2. Flexibility and Adaptability
3. Commercial Viability

Sociocultural and functional quality

1. Thermal Comfort
2. Indoor Air Quality
3. Acoustic Comfort
4. Visual Comfort
5. User Control
6. Quality of Outdoor Spaces
7. Safety and Security
8. Design for All
9. Public Access
10. Cyclist Facilities
11. Design and Urban Quality
12. Integrated Public Art

Technical quality

1. Fire Safety
2. Sound Insulation
3. Building Envelope Quality
4. Adaptability of Technical Systems
5. Cleaning and Maintenance
6. Deconstruction and Disassembly

Process quality

1. Comprehensive Project Brief
2. Integrated Design
3. Design Concept
4. Sustainability Aspects in Tender Phase
5. Documentation for Facility Management
6. Environmental Impact of Construction
7. Construction Quality Assurance
8. Systematic Commissioning

Site quality

1. Local Environment
2. Public Image and Social Conditions
3. Transport Access
4. Access to Amenities

(DGNB, 2016)

Ecological settlements

Ideas and principles of ecological town planning were presented and discussed in many publications of the late 1980s. Since the 1990s, a variety of trial and pilot projects have emerged which are especially committed to providing solutions for the ecological challenges of urban development. From 1996 to 2003, the Bundesamt für Bauwesen (Federal Office for Building and Regional Planning) launched a research area '*Stadt der Zukunft*' (City of the Future) to find

> indications of how policies of urban development could be designed in a sustainable way, and how their progress could be accompanied and evaluated using indicators. The research area 'Cities of the Future' followed up on the 1996 HABITAT II conference. It is conceptualised as a contribution to the National Sustainability Strategy as regards urban planning.
>
> (BBSR, 2015).

The internet platform www.oekosiedlungen.de, which went online in 2000 and is run by volunteers, includes an overview of over 183 settlements comprising more than 26,000 residential units in Germany. This overview has been collated over the past thirty years, but there is no claim to completeness. The goal of the website is to 'make "sustainable planning approaches" and the state of technology accessible to planners, developers and decision-makers, but also to document projects of historical significance'. By now there is a variety of online information sources. The federal website 'Sustainable Construction' (www.nachhaltigesbauen.de/)

deserves special mention. The portal covers everything from sustainable urban development and neighbourhood assessment to the assessment of construction products. The DGNB has also developed a certification scheme for sustainable neighbourhoods: 'The DGNB schemes for districts include a separate criteria set which addresses issues such as changing urban microclimate, biodiversity and interlinking habitats, and the social and functional mix' (DGNB, 2016).

Also relevant here are strategies at the European level, such as the Leipzig Charter. The position paper by the Deutscher Städtetag (Association of German Cities) on sustainable urban development (AGC, 2013) is similarly significant. The latter refers to the importance of a compact, lively, socially and culturally integrated, green city. These ideas are apparent not only in the fundamental position but also in the different strategies and actions that the paper recommends. Among the principal cross-cutting topics of strategic significance, the paper stresses city cooperation, social coherence, education, science and creativity, climate-change mitigation and adaptation, the energy shift, and the city as home. The important question is how municipalities will translate these general goals into binding targets for urban planning and construction. In this regard, every city develops its own strategy and indicators of success. Many cities elaborate their own sustainable development charters in which they attempt to represent all kinds of interest groups. The Freiburg Charter won the Academy of Urbanism's European City of the Year Award in 2010 as an 'outstanding example of sustainable urbanism'. All of Freiburg's officials signed the charter. It is a negotiated commitment and provides the framework for development and town planning in the city.

Barriers, drivers and challenges for sustainable communities

Much has been achieved with regard to technological advances towards ecologically sound design and construction, legislation and social stability. However, despite regulations and legislation, lifestyle has not changed that much towards a more sustainable way of consumption, and the trend to consume more continues. The idea of this chapter is to leave aside the discussion on sustainable development goals and strategies and to look instead at statistical or calculated figures of land and resource consumption in search of evidence of a sustainability turnaround. The calculations are based on statistical data (land use and energy consumption) and bottom-up calculations of building materials as stocks and flows for housing.

Land

The population of Germany was 69 million in 1950 and roughly 81 million in 2015 (a 17 per cent increase over 65 years). The total area used for residential, commercial and traffic (SuV) purposes has almost doubled from 1950 to the present – from 25,000 to almost 50,000 square kilometres (Einig, 2005; BBR, 2007). The hope is that in the future this increase will start to slow down.

This hope is based on the assumption that Germany will reduce its current new land consumption rate of approximately 80 hectares per day (UBA, 2003) to 30 hectares per day by 2030 (BMVBS, 2011). Many case studies reveal that even cities with shrinking populations will increase their land consumption at least a little, due to the expansion of commercial areas or the construction of an occasional small single-family home estate (Deilmann and Haug, 2010). Therefore, it remains unclear when the peak of land use for urban areas will be reached.

According to the latest census data (from 2011), the total residential floor space in Germany amounts to 3.5 billion square metres – that is, 42 square metres of residential area per capita. Subtracting vacancies nationwide, average per capita occupancy was 40 square metres of floor space in 2010. Only Denmark, Sweden, Luxembourg and the United Kingdom have more floor space than Germany. Eight European countries have less than 30 square metres per person. Our calculations assume a rise of occupied floor space by 15 per cent until 2060 (Effenberger *et al.*, 2014). At present, there are no signs that per capita floor space use will stagnate or decrease in the future.

Materials

Gruhler and Böhm (2011) indicate that the material stock of residential buildings has almost tripled since the 1950s. This is an expression of increasing prosperity,

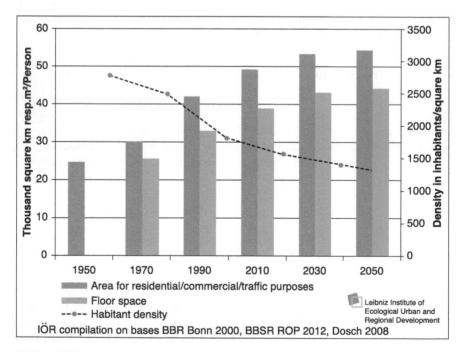

Figure 12.1 Area for SuV, floor space and inhabitant density, Germany, 1950–2050.

reflected primarily in the increase in residential space. In 2010, the total quantity of material in use in all residential buildings amounted to approximately 9.5 billion tons. At the level of the housing material stock, single- and two-family homes account for 62 per cent of the total material stock of residential floor space, a larger stock of material than that used in multi-family dwellings (38 per cent). If infrastructure is taken into account, the difference in material use between single- and multi-family houses is even greater. The average material stock for housing and internal traffic infrastructure in residential neighbourhoods was approximately 90 cubic metres per person in 2010 – 50 per cent more than in 1970.

There are numerous reasons for this growth. First, the trend towards more residential space per person is still dominant. Second, the demand for single- and two-family homes has increased greatly over time, and they are generally more material-intensive per square metre of residential floor space than multi-family buildings. Single-family homes additionally have greater specific traffic area shares. A reduction in residential density per hectare, further aggravated by population decline, and vacancies cause decreased building utilisation rates. The share of multi-family buildings, and hence the material stock of this more resource-efficient segment, has declined significantly over the last sixty years (DESTATIS, 2014).

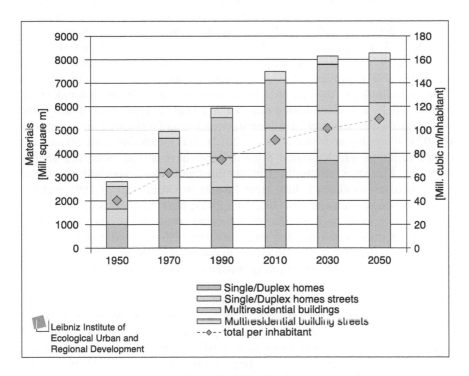

Figure 12.2 Material stock of residential buildings in Germany over time.

Energy

It is not easy to find consistent data on energy use over a long period of time. Moreover, the separation between East and West Germany up to 1989 makes an energy account even more problematic (AGE, 2009–2015). Nevertheless, primary energy use certainly increased significantly between the 1950s and the 1990s, from 150 million tons of coal equivalent (TCE) to 510 million TCE. By 2010 it had declined slightly to approximately 480 million TCE (AGE, 2009–2015). Since the 1990s, final energy used in the household sector has remained almost constant at a level of 80–85 million TCE (RWI, 2012). The main energy sources in the sector are: gas (45 per cent), oil (25 per cent), renewables (15 per cent) and others, such as coal and brown coal. Electricity consumption in the household sector rose from 13 to 18 million TCE (electric space heating included) between 1990 and 2010. Total per capita primary energy CO_2 use of the whole economy in 1990 was approximately 12 tons of CO_2 per person, which had dropped to 9.8 tons by 2010. One reason for this decline was the globalisation of markets, with the shift of energy-intensive economic segments to other countries, especially China. Moreover, the renewable energy sector grew by a factor of ten during the same period.

Some progress is being made with regard to residential space heating. According to the Energy Account Working Group, specific space heating consumption in the household sector in kWh per square metre per year dropped from 220 during the 1970s to 200 during the 1990s, and it now stands at 150 (BMWI, 2013). This equates to a drop of 25 per cent since 1990, and of 32 per cent since 1970. In terms of energy use, therefore, the housing stock has seen a considerable improvement. The principal reason for this is more efficient heating technology: thermal insulation measures have been undertaken on building shells at a rehabilitation rate of 1.7 per cent over the past fifteen years, which involves no less than 50 million square metres of residential space per year, or 600,000 housing units (Krauß *et al.*, 2012). This means that some 50 per cent of all buildings which existed prior to 1990 have already received energy upgrades. Moreover, stricter heat insulation regulations, especially the *Dritte Wärmeschutzverordnung* (Third Heat Insulation Ordinance) of 1995 and the Energy Savings Ordinances of 2002 and 2009, have contributed considerably to a reduction in average specific space heating consumption. The construction of around 8 million new housing units between 1990 and 2010 (approximately 22 per cent of the 2010 stock), which had to comply with the new ordinances, also caused average specific space heating consumption to drop by about 14 per cent (Krauß, *et al.*, 2012).

Increasing the rate of rehabilitation of the housing stock is a primary goal of the federal government's energy strategy. The large stock of single- and two-family homes, which have had either no or only slight rehabilitation, will be a significant challenge for future generations who wish to increase energy efficiency. Considerable improvement work is required for these single- and two-family homes (Simons, 2012). Concepts like low-energy houses, passive houses and/or houses made from renewable resources can be applied easily in the single-family homes segment. On the other hand, these houses generally have approximately 40 per cent more thermic shell area per square metre of residential area. This results in higher specific costs per square metre than in multi-family houses.

Housing vacancies and shortages

There are two challenges with regard to the future sustainability of communities. First, in the long run Germany's population will decline. The 2011 census recorded that there are already more than 1.8 million vacant dwellings in Germany. This is a result of demographic change and internal migration over the past twenty years. Long-term projections assume a rise in vacancies until 2060. Second, migration patterns may be decisive for the possible 'match' or 'mismatch' between demand and supply of housing. There are clear indications that younger people tend to move towards larger cities. Moreover, this trend of 're-urbanisation' and consequent housing shortages might accelerate with the arrival of more international migrants, including refugees and asylum-seekers.

Housing vacancies in East and West Germany

Although state intervention has tried to alleviate the problem of high vacancy rates, East Germany is still particularly affected by this issue. As we have seen, in Saxony and Saxony-Anhalt, two East German federal states, the vacancy rate was approximately 10 per cent in 2016, and possibly as high as 25 per cent in certain neighbourhoods. Effenberger *et al.* (2014) estimate that the vacancy rate (excluding temporary vacancies from tenancy changeovers) might rise to 11 per cent by 2060 – that is, around 800,000 vacant dwellings.

In West Germany around 1.2 million dwellings were vacant in 2016 – around 4 per cent of the total stock. The highest vacancy rate – 6.6 per cent – is found in multi-family housing built before 1918. Germany is committed to accommodating growing numbers of refugees and asylum-seekers, but given the country's predicted long-term natural population decline, it is expected that vacancy rates in West Germany will continue to rise until 2060, perhaps to 8 or even 10 per cent – that is, between 2–3 million dwellings. All scenarios indicate that vacancies will be a challenge for future development and the sustainability of communities (see Table 12.1).

Housing shortages

The period of growth in population, the economy and cities across Germany is already over. Germany is characterised by a patchwork of contraction and growth. These opposite tendencies can coexist in close spatial proximity, within a region or even within a city. This has been known to policy-makers and planners since the East German vacancy problem became a major concern. For some years now, a new phenomenon has been identified: between sixteen and eighteen big German cities with good educational opportunities and attractive social and cultural infrastructures have become powerful magnets for 20–35-year-olds. In 2015,

Professor Harald Simons coined the term '*Schwarmstädte*' (swarm cities) to describe these places (Empirica, 2015). This term is based upon Simons's observation that young people are prepared move to these cities 'at any cost', even though living space is scarce and rents are disproportionately high. Housing is expensive due to the scarcity of dwellings and land for urban development. This is a result of a restrictive policy of area designation for residential development among these cities, which is in line with the German sustainability goal of reducing daily land consumption from 80 hectares per day to 30 hectares per day by 2020. Hence, building land in these growing cities is becoming increasingly scarce. Even if estates from the 1950s, industrial wasteland or conversion areas are subjected to ex-post densification, the demand for housing will not be met (Becker, 2015). Therefore, urban planners are now discussing new upper limits for urban density. They have suggested a paradigm shift regarding the construction of new high-rise developments. In Germany, the latter have been discredited since the 1970s due to social concerns. These discussions about densification are inconsistent with the aims of a socially compatible green city adapted to the climate, the protection of flora and fauna and emission control. For example, the city of Freiburg, one of the swarm cities and mentioned above as an example of sustainable urbanism, has limited building height to four storeys in its sustainability charter after discussions about resilient and sustainable residential areas. Now, though, it is confronted with the challenge of moderating rapid rent increases and the possibility of surrendering its potential for urban development to the regional outskirts, accepting more commuters and putting up with a higher traffic load.

The current volume of refugees and asylum-seekers is a further challenge for Germany and city housing. The cooperative and municipal housing sectors may have to accommodate a large share of the refugees. The big housing estates of the 1970s already accommodate tenants and migrants on low incomes. There is a likelihood that new refugees will migrate to these areas, too. Over the last two decades, these housing organisations have gained a lot of experience with integration and how to avoid stigmatisation. The government's '*Soziale Stadt*' (Social City) programme financially supported '*Quartiersmanagement*' (neighbourhood management). The question is now how to continue with this programme and how to prevent overconcentration of low-skilled and poorly integrated newcomers in the big housing estates. The GdW Bundesverband deutscher Wohnungs- und Immobilienunternehmen (Union of Real Estate Companies) argues that a sensible and spatially distributed assignment of refugees is necessary to avoid conflict with the existing inhabitants (GdW WI, 2016). This has to be done through voluntary cooperation between municipalities and the housing organisations. Also, a long-term programme to provide financial support to integration efforts is required and the government has to think about incentives for the construction of new social housing (GdW WI, 2016). As we have seen, in January 2016, the Federal Ministry for the Environment, Nature Conservation, Building and Nuclear Safety (BMUB) doubled the aid for social housing to €2 billion, and tax incentives for new construction are now under discussion.

Conclusions

Over recent decades, much has been achieved regarding sustainable and environmentally compatible urban development. As a result, living conditions in Germany's cities have clearly improved in terms of environment and health. A large number of projects confirm this. Nevertheless, when it comes to large urban development plans and projects like motorways and airports or the protection of big industries, the conflicts between ecological and economic interests persist and are still serious. Despite the fact that sustainability is now enshrined in the German constitution, Rehbinder (2015: 257) highlights the problem of operationalising sustainability in the German context, noting that 'considering the actual effectiveness of the legal rule on sustainability, the findings are sobering'. He states that the requirement to balance ecological, economic and social aspects continues to make any assessment difficult. For example, it is difficult to assess the economic consequences and distributive effects, and the regenerative ability and ability to replace natural resources are questionable. On that basis, there are always loopholes that hinder agreement on environmental goals, especially if the latter may result in financial losses. According to Rehbinder (2015: 257–268), it is necessary to drop the balancing paradigm or to modify it towards the 'ecological rule of law', which amounts to recognising nature in its own right. Nevertheless, the success of ecological urban planning, construction and housing, as well as the role of indicator-based monitoring and supporting certification schemes (see DGNB, 2016), should not be underestimated.

At the moment, questions pertaining to demographics, internal migration and immigration from abroad are displacing the discussion about ecological issues of sustainable urban development. After a peak in housing demand in five years' time, the main demographic issue will be population decline. In the long run, Germany will shrink from 80 million to 70 million inhabitants. Whether this will happen in 2060 or 2070 remains to be seen. Net annual in-migration of over 500,000 could halt this tendency. In 2015, annual immigration was 1.5 million people. At the same time, on average around 650,000 inhabitants leave Germany every year, although in 2014 it was 900,000. If net in-migration of 500,000 per year were to continue for twenty years, this might strain society to breaking point. The long-term population projections of Germany's Statistical Office assume an annual in-migration of between 150,000 and 250,000. In these projections, from 2025 onwards, Germany will lose 250,000 to 350,000 inhabitants every year. In any case, Germany and Europe will be much different from today in ten years' time: 'The 21 century will be the century when we have to share our wealth with those whom we exploited before' (Precht, 2015). Refugees come to Europe, and especially to Germany, to escape from war, but many also come to participate in 'winning societies'. Integration and social inclusion will demand billions of euros for education, coaching and caring in the near future. Integration and the participation of welfare recipients in societal life will be one of the big challenges for the sustainability of communities over the next few years.

Despite the problems mentioned above, the overall objective of sustainable development and its implementation in sustainable neighbourhoods will not be questioned. Urban planning has included ecological concerns in its corpus of rules and anchored them legally. The living and environmental qualities of German cities are good and will not be undermined by temporary challenges. However, there is another aspect that affects us personally: there is no sign that citizens might soon reduce their consumption levels. Just looking at the residential purposes, land use for SuV and material use for housing per person have increased due to growing floor space and lower SuV density. Thermal comfort has also increased. Luckily, due to new technology and better building energy performance, energy consumption per person has decreased. These tendencies – more space, more materials – are unlikely to change in the near future. Wealth has increased year on year. Consequently, the question now being asked is whether, in addition to environmental protection and efficiency goals, it is necessary to think critically about our lifestyles and consumption patterns. Only if we seriously consider sufficiency strategies might we discover a more sustainable lifestyle that allows for wealth while using less land, energy and materials.

References

AGC (2013) *Integrated Urban Development Planning and Urban Development Management: Strategies and Instruments for Sustainable Urban Development.* Position paper. Berlin and Cologne: Association of German Cities.

Arbeitsgemeinschaft Energiebilanzen (2009–2015) *Energieverbrauch in Deutschland.* Available at www.ag-energiebilanzen.de/ (accessed 02.02.2016).

BBR (2007) *Regionale Siedlungsflächenentwicklung in den neuen Bundesländern auf Basis von Prognosen der Bau-und Immobilienwirtschaft.* Bonn: Bundesamt für Bauwesen und Raumordnung.

BBSR (2015) *Forschung im Blick 2015/2016.* Available at www.bbsr.bund.de/BBSR/DE/Veroeffentlichungen/Sonderveroeffentlichungen/2015/Forschung-im-Blick-15-16.html (accessed 13.07.2016).

Becker, C. (2015) *Neue Dichte in der Stadt – Das komplexe Verhältnis von Stadtraum, Dichte und Bevölkerungswachstum* [*New Urban Density: The Complex Relationship between Urban Space, Density and Population Growth*]. Heuer Dialog INSIGHT No. 1, March.

BMVBS (2011) *30 ha Ziel realisiert – Konsequenzen des Szenarios Flächenverbrauchsreduktion auf 30 ha im Jahr 2020 für die Siedlungsentwicklung.* Berlin: Bundesministerium für Verkehr, Bau und Stadtentwicklung.

BMVBW (2001) *Leitfaden Nachhaltiges Bauen.* Berlin: Bundesministerium für Verkehr, Bau und Wohnungswesen.

BMWI (2013) *Gesamtausgabe der Energiedaten – Datensammlung des BMWi. Letzte Aktualisierung: 12.01.2016.* Available at www.bmwi.de (accessed 29.01.2016).

Cischinsky, H. (2013) 'Bedeutsam, aber unbekannt – der Privatvermieter', in *IWU-Jahresberichte 2013*. Darmstadt: IWU.

Deilmann, C. and Haug, P. (eds) (2010) *Demographischer Wandel und technische Infrastruktur: Wer soll die Kosten tragen? Eine Untersuchung am Beispiel ostdeutscher Mittelstädte.* Aachen: Shaker.

DESTATIS (2012) *Bevölkerung und Erwerbstätigkeit. Haushalte und Familien*. Wiesbaden: Statistisches Bundesamt.

DESTATIS (2013) *GWZ 2011 – erstellte Auswertungen des Statistischen Bundesamtes*. Wiesbaden: Statistisches Bundesamt.

DESTATIS (2014) *Zensus 2011: 0,5 Millionen Wohnungen mehr*. Wiesbaden: Statistisches Bundesamt.

DESTATIS (2016) Various statistics. Available at www.destatis.de (accessed 01.02.2016).

DGNB (2016) *Zertifizierungssystem*. Available at www.dgnb-system.de/de/system/ zertifizierungssystem/ (accessed 02.02.2016).

Effenberger, K.-H., Banse, J. and Oertel, H. (2014) *Deutschland 2060 – die Auswirkungen des demographischen Wandels auf den Wohnungsbestand*. Stuttgart: Fraunhofer IRB-Verlag.

Einig, K. (2005) 'Demografischer Wandel, Siedlungsentwicklung und Infrastrukturkosten'. Available at www.staedtestatistik.de/fileadmin/vdst/Braunschweig2005/BS330_DGD. pdf (accessed 8.8.2016).

Egner, B., Georgakis, N., Heinelt, H. and Bartholomäi, R. C. (2004) *Wohnungspolitik in Deutschland. Positionen. Akteure. Instrumente*. Darmstadt: Schader-Stiftung.

Empirica (2015) *Schwarmstädte in Deutschland – Ursachen und Nachhaltigkeit der neuen Wanderungsmuster in Deutschland. Ausgewählte Ergebnisse*. Available at http://web. gdw.de/uploads/pdf/Pressemeldungen/2015001_GdW_Schwarmstaedte_Ergebnisse_ endg.pdf (accessed 18.01.2016).

GdW WI (2016) *Großsiedlungen – Integrationsaufgaben müssen im Mittelpunkt der Zuwanderung stehen*. Freiburg: Haufe-Lexware GmbH.

Gruhler, K. und Böhm, R. (2011) *Auswirkungen des demografischen Wandels auf das Stofflager und die Stoffflüsse des Wohngebäudebestandes – Deutschland 2050 [Effects of Demographic Change on the Material Stock and Material Flow in the Residential Housing Stock – Germany 2050]*. Stuttgart: Fraunhofer IRB-Verlag.

IVV (2015) 'Wohngeld – soziale Strukturen und Wirkungen', *Immobilien Vermieten und Verwalten*, October.

Korte, H. (1972) *Soziologie der Stadt, Grundlagen der Soziologie*. München: Juventa Verlag.

Krauß, N., Deilmann, C. and Gruhler, K. (2012) 'Wo steht der deutsche Gebäudebestand energetisch? Modernisierungsstand, Ausgangsbasis und Perspektiven', *Kurzberichte aus der Bauforschung* 53(5): S40–S50.

Krusche, P., Althaus, D. and Gabriel, I. (1982) *Ökologisches Bauen*. Wiesbaden: Bauverlag GmbH.

Meadows, D. (1972) *Die Grenzen des Wachstums*. Hamburg: Deutsche Verlags-Anstalt.

Precht, R. D. (2015) 'Echte Träume – echte Not', *Die Zeit*, 30 December.

Rehbinder, E. (2015) 'Neue Nachhaltigkeitskonzepte im Umwelt- und Planungsrecht - ein Beitrag zu einer nachhaltigen Wirtschaft und Gesellschaft?', *Zeitschrift für Umweltpolitik and* Umweltrecht 38(3): S257–S268.

RWI (2012) *Erstellung der Anwenderbilanzen 2010 und 2011 für den Sektor Private Haushalte*. Essen: Endbericht des Forschungsprojekts im Auftrag der AG Energiebilanzen e.V.

Schumacher, E. F. (1978) *Die Rückkehr zum menschlichen Maß*. Reinbek: Rowohlt.

Simons, H. (2012) *Energetische Sanierung von Ein- und Zweifamilienhäusern – Energetischer Zustand, Sanierungsfortschritte und politische Instrumente*. Berlin: Empirica.

Spiegel, S. (1978) *Wohnungsbau in der BRD – Eine Dokumentation der Wohnungspolitik und ihrer Ergebnisse*. Aachen: RWTH Aachen, Lehrstuhl und Institut für Wohnbau.

Statista (2014) *Struktur der Raumwärmebereitstellung in privaten Haushalten in Deutschlandnach Energieträger im Jahr 2014.* Available at http://de.statista.com/statistik/daten/studie/250403/umfrage/raumwaermebereitstellung-nach-energietraeger-in-deutschen-haushalten/ (accessed 27.01.2016)

UBA (2003) *Reduzierung der Flächeninanspruchnahme durch Siedlung und Verkehr – Materialienband.* Available at www.umweltdaten.de/publikation/fpdf-1/2587pdf (accessed 13.07.2016).

Vale, R. and Vale, B. (1975) *The Autonomous House.* London: Thames and Hudson.

13 Conclusions

Nessa Winston and
Montserrat Pareja-Eastaway

Cities and towns are facing a wide range of very significant economic, social and environmental challenges. Climate change and the effects of the GFC have increased the severity of these challenges. The issues which require appropriate responses include: debt (sovereign, public and personal); unemployment, underemployment and precarious employment; aging populations; increasing diversity, migration and the refugee crisis; environmental problems, such as pollution and waste; climate change (e.g. GHG emissions, energy poverty and security, flooding and drought). Sustainable urban communities are more important than ever as a means of addressing some of these problems, and housing and urban planning can play important roles in contributing to their development. However, in many cases there are inadequate resources and/or political will to address them and ensure the welfare of present and future generations as well as the natural environment.

This volume has explored the sustainability of urban housing and communities in eleven European countries, both EU and non-EU member states. The countries examined reflect the diversity that exists within this geographical region in terms of a number of important, relevant considerations, including housing systems, welfare regimes and environmental performance. Some countries are generally considered to be 'leaders' in terms of sustainable development (e.g. Germany, Denmark and Sweden), commencing work in this area before other nations and making considerable progress in certain respects, particularly on the environmental pillar. Others have developed strategies at a later stage, linked to pressure from international organisations (e.g. UN and the Kyoto Protocol and/or EU accession and/or membership) and progress has been much more limited to date. Thus, they may be viewed as being at different stages along a sustainable development continuum.

Several definitions, policies and indicators relating to sustainable development and sustainable communities have been utilised in the countries examined, at both national and local levels. The Brundtland Commission's definition of sustainable development (WCED, 1987), with its emphasis on social, economic and environmental pillars, is evident in almost all countries. Similarly, most countries have SD strategies and Local Agenda/Action 21 activities which may be linked back to the 1992 UN Earth Summit in Rio and subsequent UN conferences.

In some – but not all – countries, the term 'sustainable community' is employed and linked to the EU's Bristol Accord definition. In other countries, in particular Romania and Hungary, less work has been done in these areas, and where it has been done it was as a prerequisite for accession to the EU and/or its funding mechanisms. In some 'leader' countries, there has been a shift from a broader view of SD to a narrower focus on the environmental dimension (e.g. Sweden), especially energy, or on economic growth, innovation and competition (e.g. the Netherlands).

Challenges for sustainable communities and urban housing in Europe

The case study chapters reveal variations in the extent to which sustainable communities are valued, and in the priorities given to environmental, social and economic issues and goals. The GFC might be viewed as a turning point for many countries, regardless of differences in the magnitude of its consequences, as progress towards sustainable communities has slowed or stalled and government priorities changed in many cases. Cutbacks in public budgets and austerity measures are important aspects of the explanation for this. Sustainable communities are about the 'local', but they depend on national and/or regional policies. In addition, the institutional thickness in a country contributes to the complexity of defining and achieving goals, regardless of its financial circumstances.

Among the many challenges to sustainable housing highlighted in the preceding chapters, some emerge as particularly problematic in most countries. These include housing affordability, a mismatch between housing supply and demand, high levels of residential energy consumption and associated emissions, and energy poverty. The problem of housing affordability is probably the most common one facing these countries. It affects those on lower incomes the most, usually younger people, the unemployed and (im)migrants. However, in some cities and regions, those on middle incomes also experience significant difficulties due to high housing costs in some locations. Housing cost issues arise across all housing and welfare systems. It is important to note that housing affordability is a crucial, but at times neglected, dimension of sustainable housing and communities.

Significant effects on the urban environment arise from the relationship between housing costs and household income. These range from gentrification, social segregation and exclusion to overcrowding, slum conditions and homelessness. A related concern in a number of locations is the mismatch between housing need and supply. Several examples of these effects can be seen across the chapters of this book. A number of countries have regions with shrinking cities, high levels of housing vacancies and yet strong demand, and/or populations concentrated in a small number of areas (e.g. Ireland, Spain, Germany, Hungary, the UK and Sweden). These issues may arise for different reasons (e.g. problems in planning and regional economic development) and, therefore, require different policy responses. In a similar vein, the provision of social housing is fundamental in ensuring shelter. There is a diversity of definitions and conceptualisations of it across European countries, from universal access to a narrow, targeted approach. However, social housing can play key roles in each country's mechanisms to guarantee housing access and in its achievements regarding sustainable communities.

High levels of residential energy consumption and associated emissions are a considerable problem in many of the countries examined. This includes countries with advanced building regulations and a housing stock that should facilitate low energy consumption (e.g. Denmark, Germany, Sweden and the Netherlands). However, rising consumption may be due to a combination of 'rebound effects' and affluence (increased dwelling size, second homes, ever more electrical devices and so on). Many other countries have initiated programmes to improve the energy efficiency of their stock via building regulations and/retrofitting. Some of these reveal a decline in consumption but this may be a result of the effects of the recession on the financial situation of households, and consumption may rise again following the recovery.

Another common theme in the chapters is that of tensions between the economic, social and environmental pillars of sustainable development. The most prominent examples relate to conflict between the economic and environmental dimensions. For example, in some countries the construction industry has attempted to reduce or complain about energy regulations and space standards on the grounds of construction costs (e.g. Sweden, Ireland and the UK). In addition, attempts to increase residential density and urban compaction on environmental grounds may reduce the quality of life of residents when not done well, revealing tensions between social and environmental pillars. There are also many examples where the focus is on one pillar to the neglect of others, as in Denmark, where policies now target energy and climate change. However, usually the economic pillar is prioritised by many relevant actors (including households) over the social and environmental pillars, perhaps unsurprisingly given the impact of the GFC on many of the countries examined here. The extent of the frustration with this, even in countries with long-standing commitment to environmental issues, such as Germany, is evident from the fact that the authors of the German chapter, Deilmann and Effenberger, ask whether it is time to move beyond 'the balancing paradigm' and give more emphasis to the ecological pillar and to climate change. While inadequate attention to environmental issues creates and/or exacerbates existing economic and social problems (e.g. health, quality of life, flooding) and intensifies climate change effects, if social pillar issues are neglected, some policy changes may result in more social exclusion. This means that 'poverty proofing' – examining the effects of policy changes on income distribution – is essential prior to introducing or changing policies. For example, fuel and transport poverty might be alleviated by providing good, affordable public transport prior to increasing carbon taxes.

Towards more sustainable communities

Good policy and its implementation

The case study chapters outlined numerous barriers to progress on more sustainable urban housing. These present very considerable challenges and are not easily overcome. What is clear is that both good policy and its implementation are essential for the sustainable regeneration of housing and communities, addressing each of the social, economic and environmental pillars. However, problems in practice with mutually reinforcing pillars are clear in some of the countries, including those that have made progress on environmental issues but where social and/economic issues remain or have become more challenging. Designing and implementing a successful local sustainable community policy/vision is a collective process, a partnership between relevant stakeholders, including residents. It means ensuring that variations in local needs are addressed, such as those of low-income groups, older people and members of minority ethnic groups.

In the context of public funding restrictions and limitations on borrowing faced by members of the Economic and Monetary Union, it is necessary to examine and utilise alternative methods of financing housing (both social and affordable) and the necessary physical, social and environmental infrastructure required for sustainable urban neighbourhoods. This could include European Investment Bank loan finance, which offers low-cost, long-term finance to housing associations for social housing construction, with the value of the loan matched by the national Housing Finance Agency, as has happened in the UK and Ireland, for example. Similar loans are available for energy retrofitting of social housing and community-based energy renewal systems. Another approach involves 'tying' EU funding to measures that combat social segregation, an approach which appears to have been effective in Hungary. Given the shortage of affordable housing in urban areas, it will be important to resist any EU interventions, on the grounds of rules on competition and state aid, to restrict social housing in some countries. Gruis and Elsinga (2014) suggest that interventions to date in a small number of countries should not be exaggerated and that national governments must retain the primary role in relation to social housing. Housing Europe (the European Federation of Public, Cooperative and Social Housing) and others have been campaigning on this issue, with positive responses so far from the European Commission (see, e.g., Housing Europe, 2016; Delli, 2012; Hencks, 2012).

Policies to promote the development of sustainable urban communities need to focus on: locations where people want to live and/or regenerating areas so that they are attractive to residents; recognising the intrinsic value of the natural environment as well as its role in enhancing the quality of life of residents; leisure; renewable energy; food production; biodiversity; air quality; water (waste and flood) management; and reducing heat stress. That is, green/blue–green infrastructure and building with nature can have a broad range of positive impacts, not just in relation to climate adaptation and mitigation. However, such policies are often absent and/or insufficient in themselves, because when they are drafted they are frequently not implemented due to some of the barriers detailed in this volume.

In light of their extensive experience of working on sustainable urban neighbourhoods, Rudlin and Falk (2009: 296) argue that past failures were due to an over-focus on the 'end state'. They suggest a phased strategy in stages, and, while sticking to the key aspects of the plan, maintaining a degree of flexibility on details, given the unpredictable nature of the future. Related ideas are expressed in the Norwegian chapter, drawing on Sennett's (2013) concept of the 'open city'. However, Rudlin and Falk (2009) also argue that past failures can be attributed to the inadequate resourcing of plans, a situation that is relevant for a number of the countries in this volume. They argue that investment plans must be part of the policy, that it is important to target areas that will have the most impact, and that under-utilised infrastructure must be used (Rudlin and Falk, 2009). In relation to resources, it is important that all relevant stakeholders, including residents, are provided with the necessary assets to participate adequately in the design of these plans, a point to which we return below.

Governance: involving relevant stakeholders

Some definitions of 'sustainable development' include a 'governance/institutional' pillar as well as the environmental, social and economic pillars (see, e.g., Pareja-Eastaway and Støa, 2004). A number of the chapters in this volume have highlighted problems relating to transnational, national, regional, city and more local-level governance. These include issues relating to: national government policy priorities (e.g. changes in national government policy priorities and planning gain in the UK; the discontinuation of the development of micro-regions and the social proofing of development plans in Hungary); lack of trust impeding collaboration among relevant local stakeholders (collaboration on complementary sustainable community initiatives such as housing, energy retrofit, district heating and sustainable transportation in Bucharest); decentralisation of land use decisions to extremely small municipalities (urban sprawl in Budapest); and localised planning impeding housing supply in high-demand areas (e.g. the UK and Ireland).

Sustainable community development requires leadership and good governance. There are a number of signs of leadership, including the Covenant of Mayors for Climate Change and Energy, consisting of local and regional authorities who have voluntarily committed to implementing EU climate and energy objectives in their territories. In addition, the Declaration by the Mayors of the Capital Cities of the European Union on the EU Urban Agenda and the Refugee Crisis (2016) advocates social and affordable housing for a range of citizens in liveable neighbourhoods. Finally, leadership changes in some European cities (e.g. Prague, Madrid, Barcelona and many Polish cities) have resulted in the prioritisation of social and environmental sustainability in urban development. However, similar commitments are required at national and regional level if more progress is to be made on sustainable urban housing development.

Implementing more sustainable urban communities requires action from a range of stakeholders at a variety of levels. These include those highlighted by Baker (2015) in her vision of more sustainable development, namely: international diplomacy and governance (UN and EU); national and sub-national governments; civil society and economic actors; and individual values and behavioural change. From a housing perspective, the key national and local stakeholders include planners, the construction industry, architects, estate agents, housing managers (housing associations, municipalities), utility companies and households. Many significant challenges identified in this volume relate to these important stakeholders. For progress to be made on sustainable communities, each of them needs to sign up to a sustainable communities policy, which might entail the use of 'charters' or 'protocols' such as that used in Freiburg (described in the German chapter) or other sustainable urban neighbourhoods and eco-villages (Rudlin and Falk, 2009; Winston, 2012). These might include a number of shared principles to which stakeholders commit when signing the charter.

Broad partnership approaches involving all of the relevant actors (e.g. planners, architects, the construction sector and residents) are important but not easy to negotiate; moreover, they do not always result in solutions that are environmentally sustainable. As is highlighted in the Norwegian chapter, promoting more sustainable lifestyles 'requires a radical reconfiguration of values and consumption patterns', changing the dream from a detached house with a private garden and the need to 'push the limit of what developers and real estate agents consider the market wants, or what is economically feasible' (Chapter 5, this volume), otherwise the necessary changes in individual values and practices will not occur. The design, regeneration and implementation of sustainable communities could be facilitated by professionals based in regional or national centres of excellence, as most municipalities are too small to employ all of the necessary expertise (Rudlin and Falk, 2009; Winston, 2012). These experts could work with local communities to design or regenerate neighbourhoods so that they are more economically, socially and environmentally sustainable. This would mean supplying residents with the necessary resources, presenting them with a range of sustainable alternatives, examples of where high-quality urban living has been achieved without having to reside in a house with a front and back garden. Communities are not homogeneous and views will differ, but 'planning for real' approaches can be useful in cases where there are differences of opinion (Winston, 2012). Work of this type could include educating residents about the challenges associated with sustainable communities, including rising consumption levels linked to housing and lifestyles, and exploring ways to reduce consumption. This broad partnership approach would, of course, be complemented by 'top-down' policies, such as building regulations, and 'bottom-up' approaches, such as eco-villages established by small groups.

Range of policies and instruments

The evidence presented in this volume suggests that the best results might be achieved by using a broad range of policy tools and targeting them at the relevant actors. These can include the following types of instruments: regulatory (e.g. building regulations); market based (e.g. carbon taxes); informational (e.g. labelling and certification schemes such as DGNB in Germany or Minergie in Switzerland); and education/training for relevant professionals and residents. There may be financial disincentives, such as taxes on carbon or vacant sites and dwellings, or financial incentives, such as retrofitting grants, renewable energy rebates or subsidies. The Swedish chapter presented evidence that enabling residents of multi-family households to control and conserve energy can generate more positive results than centrally controlled systems. Other incentives may involve 'tying' funding to the achievement of particular goals, such as 'social proofing' to combat social segregation in Hungary. Regarding measures such as carbon taxes, a thorough assessment of their potential social impact is essential prior to implementation. As several chapters in this volume have highlighted, such taxes can be regressive, disproportionately affecting low-income households and contributing to or exacerbating existing inequalities.

A range of approaches to increase the supply of affordable housing should also be considered. In some locations, there is scope to do this by increasing housing density in inner and outer suburbs as well as some city centres without compromising quality of life (Rudlin and Falk, 2009). This entails some densification while keeping enough green space for recreation and leisure as well as food and energy production, but also, as Rudlin and Falk (2009) point out, past experience reveals that a good-quality public realm and community facilities are crucial to making higher-density living an attractive option. Overall, these are not easy tasks for a number of reasons, in particular the decline in public subsidies for social and affordable housing, financial issues in the construction sector, and local opposition to densification, especially for social housing. Various mechanisms might be considered, depending on the housing and planning context. These might consist of a range of inclusionary housing measures (see de Kam *et al.*, 2014). One example of this involves recapturing high land values via planning obligations. Section 106 in England offers such an example, which, despite the downturn and reduced housing construction, has continued to provide some affordable housing as the discretionary nature of the land use planning system has facilitated negotiations between local authorities and developers on the amount of tax paid (Morrison and Burgess, 2014; de Kam *et al.*, 2014). The importance of inclusionary housing goals means local authorities may have to accept some short-term losses in affordable housing provision (Morrison and Burgess, 2014). In this volume, Bramley points out that it requires consistent support from the national government. Others highlight that its effectiveness depends on the enforceability of the policy; a mature, affordable housing sector; and particular market conditions (Austin *et al.*, 2014).

Finally, in some locations, the supply of more affordable housing might be expanded via housing cooperatives and/or community ownership schemes. The Swiss chapter outlines how these have been used successfully in Zurich. In their work on sustainable urban neighbourhoods, Rudlin and Falk (2009) highlight the potential role of community investment schemes, community land trusts, and development trusts. The chapter on the Netherlands describes some of the potential limitations of one community enterprise project in the short–medium term, but the success of similar projects has been demonstrated in other locations over the longer term (Bailey, 2012; Hopkins, 2008; James and Lahti, 2004).

Sustainable communities require good policy not only on housing but also in related areas such as public transportation and energy. In addition, problems of housing affordability might be alleviated in some countries and regions by enhanced regional economic planning that focuses on increasing employment in declining urban areas and building up city regions that are well connected by good public transport. This would have multiple advantages, such as utilising vacant dwellings and brownfield sites and increasing the viability of public transport, and could be linked to employment strategies that emphasise green jobs/eco-industry (Angelov and Johansson, 2011; Power and Katz, 2016).

Concluding remarks

Many of the chapters in this volume have outlined inequalities by age, income and ethnicity in access to affordable housing and in housing-related consumption. Furthermore, in most high-income countries, consumption appears to be rising. In addition, while the focus here is on one part of the Global North, there are obviously significant inequalities in income and wealth between the Global North and South which are beyond the scope of this volume. A continuation of the lifestyles and consumption patterns outlined among some groups in the Global North, along with a failure to tackle climate change and other environmental issues, will exacerbate these inequalities and the effects of climate change on the Global South. In sum, they raise serious concerns about social and environmental justice both within and across these regions.

Housing is high on the political agenda for many countries due to an increased demand for affordable housing arising from a range of factors, including inadequate investment, the effects of the GFC, declining household size, and increasing migration. The role of housing as a driver of more sustainable communities could be enhanced in many countries and it could play a central role in longer-term strategies to achieve greater sustainable development overall. Furthermore, a failure to tackle the issue of inadequate housing supply and access to affordable housing will exacerbate the xenophobia that is already evident in some of the countries covered in this book, including negative attitudes towards refugees from the Global South and countries devastated by war in other regions.

Finally, there are important and useful synergies between sustainable community development policy and other relevant and important policies, including those relating to climate change mitigation/adaptation; social inclusion/ cohesion; and regional development. A combination of the three approaches could enhance the resilience of communities, socially, economically and environmentally.

References

Angelov, M. and Johansson, N. (2011) 'Green jobs', in T. Fitzpatrick, (ed.) *Understanding the Environment and Social Policy*. Bristol: Policy Press, pp. 254–270.

Austin, P., Gurran, N. and Whitehead, C. (2014) 'Planning and affordable housing in Australia, New Zealand and England: common culture; different mechanisms', *Journal of Housing and the Built Environment*, 29: 455–472.

Bailey, N. (2012) 'The role, organisation, and contribution of community enterprise to urban regeneration policy in the UK', *Progress in Planning*, 77: 1–35.

Baker, S. (2015) *Sustainable Development*. Second edition. London: Routledge.

de Kam, G., Needham, B. and Builtelaar, E. (2014) 'The embeddedness of inclusionary housing in planning and housing systems: insights from an international comparison', *Journal of Housing and the Built Environment*, 29: 389–402.

Delli, K. (2012) *Report on Social Housing in the European Union* (2012/2293(INI)). Brussels: Committee on Employment and Social Affairs, European Parliament.

Gruis, V. and Elsinga, M. (2014) 'Tensions between social housing and EU market regulations', *European State Aid Law Quarterly*, 3(14): 463–469.

Hencks, R. (2012) 'TEN/484 Social Housing as a Service of General Economic Interest. Opinion of the European Economic and Social Committee on "Issues with defining social housing as a service of general economic interest' (own-initiative opinion)'. *Official Journal of the European Union*, 2013/C 49/09: 53–58.

Hopkins, R. (2008) *The Transition Handbook: From Oil Dependency to Local Resilience*. Cambridge: Green Books.

Housing Europe (2016) 'Better EU rules for better services of general interest in housing: Housing Europe letter to Commissioner for Competition, Margrethe Vestager'. Position paper. Available at www.housingeurope.eu/resource-657/better-eu-rules-for-better-services-of-general-interest-in-housing, accessed 14 July 2016.

James, S. and Lahti, T. (2004) *The Natural Step for Communities: How Cities and Towns Can Change to Sustainable Practices*. Gabriola Island, BC: New Society.

Mayors of the Capital Cities of the European Union (2016) *Declaration by the Mayors of the EU Capital Cities on the EU Urban Agenda and the Refugee Crisis*. Available at: http://urbanagendaforthe.eu/wp-content/uploads/2016/05/Declaration_capitals_mayors_meeting_2016_amsterdam_20160420.pdf, accessed 8 August 2016.

Morrison, N. and Burgess G. (2014) 'Inclusionary housing policy in England: the impact of the downturn on the delivery of affordable housing through Section 106', *Journal of Housing and the Built Environment*, 29: 423–438.

Pareja-Eastaway, M. and Støa, E. (2004). 'Dimensions of housing and urban sustainability', *Journal of Housing and the Built Environment*, 19: 1–5.

Power, A. and Katz, B. (2016) *Cities for a Small Continent: International Handbook of City Recovery*. Bristol: Policy Press.

Rudlin, D. and Falk, N. (2009) *Sustainable Urban Neighbourhood: Building the 21st Century Home*. Second edition. Oxford: Elsevier Architectural Press.

Sennett, R. (2013) *The Open City*. Available at www.richardsennett.com/site/senn/templates/general2.aspx?pageid=38&cc=gb, accessed 14 July 2016.

Winston, N. (2012) 'Sustainable housing: a case study of the Cloughjordan eco-village', in A. Davies (ed). *Enterprising Communities: Grassroots Sustainability Innovations*. Bingley: Emerald, pp. 85–103.

World Commission on Environment and Development (WCED) (1987) *Our Common Future*. Oxford: Oxford University Press.

Index